All About Satellite TV

Karl Fincke
Technical Editor
Satellite TV Week®

Fortuna Communications Corporation
Fortuna, California

Printed in the United States of America by Humboldt Printing.

Satellite TV Week is published weekly by
Fortuna Communications Corporation,
P.O. Box 308, Fortuna, Calif. 95540-0308.

Publisher: Patrick O'Dell

Editor: James E. Scott

Technical Editor: Karl Fincke

Cover Design: Christopher Stroud
Book Design and Production: Diana Christine Bruck

ISBN: 1-879804-01-8

Table of Contents

About This Book

You are reading *All About Satellite TV* because a satellite dish system probably is the most technically advanced and complex possession most home owners will ever buy. Its technology includes or is made possible by all known forms of consumer electronics, from radio to video to computers and telephones.

Then there are some pretty sophisticated manufacturing techniques involved in producing the precision equipment that is necessary to receive signals from 22,300 miles above the equator. A dish system is far beyond a pair of rabbit ears atop the TV.

Let's not forget the encryption and decryption equipment, all of which uses the DES algorithm that is used by the federal government and military to protect secrets.

As if all this weren't enough, being a dish owner also means having an appreciation for, if not an understanding of, the high-tech world of rocket science, satellites and outer space.

Since satellite TV originally was created to deliver programming to cable companies, being a dish owner means having your own private, customized cable company. This means you are responsible for maintaining, replacing and updating equipment. That may mean you do your own work. But because satellite TV requires specialized expertise and training as well as knowledge of electricity and building codes, it probably means you have the work done by a professional, your local satellite dealer.

However, sometimes a dealer isn't available or the owner's manual doesn't cover the subject. That's why *All About Satellite TV* was written. In a handy, concise volume, it presents a detailed explanation of the basic elements of satellite TV in language anyone can understand.

For more than 11 years, one of the most dependable sources of information about home dish systems has been *Satellite TV Week.* Technical editor Karl Fincke's weekly column, "Ask the Tech Editor," is a valuable resource of technical information unavailable anywhere else.

In this book, Fincke presents a wealth of information that will help anyone who has or plans to have a home dish system. Whether you simply want to understand satellite terms or find where to get parts and repair for your old receiver, it's in *All About Satellite TV.* Enjoy!

James E. Scott
Editor

Courtesy of Hughes Communications

CHAPTER 1

An Introduction to Satellite Basics

The purpose of this book is to give you, the satellite television viewer, a basic understanding of how your system works. It will describe the function of each component, give tips on troubleshooting common system problems and suggest preventative measures you can take to ensure the best performance from your satellite system for years to come. It also will explore the latest in equipment and accessories such as surround sound processors, single channel per carrier (SCPC) receivers, C/Ku-band feeds, surge and lightning protectors, teletext receivers, pay-per-view (PPV) modems, signal combiners and much more.

Basic Systems

Satellite communication circuits are composed of three basic elements: an uplink facility that beams the signals to an orbiting satellite; a telecommunications satellite that receives the signals and transmits them back to earth; and a receiving station.

Signals that contain the images and sound which originated in the studio are sent to an uplink site where they are added onto a microwave carrier. This processes is called modulation. An extremely large uplink dish concentrates the outgoing signals and beams them up to a satellite parked approximately 22,300 miles above the equator.

The satellite's receiving antenna captures the incoming signals and sends them to a receiver for processing. These processed signals which contain the original picture and sound (modulated) information are then routed to an amplifier. This receiver/transmitter package is called a transponder. The outgoing signals are coupled to the transmitting antenna array which focuses the microwaves into a narrow beam of energy that is directed toward the earth. The specific area of the beam's coverage on the surface of the earth is called a footprint.

Part of the signal processing that takes place on board the spacecraft's transponder includes converting the uplink carrier wave to a lower downlink microwave frequency. Signals are uplinked at a frequency of approximately 6 billion cycles per second, abbreviated 6 gigahertz (GHz). The downlink frequency for 24-transponder C-band satellites ranges from 3.7 GHz to 4.2 GHz. Signals are downlinked on a slightly lower frequency to avoid interference.

Polarization

In order to get the most efficient use of the limited spectrum of frequencies allocated for satellite communications, two sets of signals share the same frequency band simultaneously. The satellite's transmitting antenna sends opposite senses of polarity. Microwave energy sent by the satellite's transmitting antennas is positioned in either a vertical (straight up and down) or horizontal (lying flat) plane. All North American C-band satellites use this linear polarization scheme, whereby half of the transponders are vertically polarized and half of the transponders are horizontally polarized. Many international satellites transmit a spiral pattern called circular polarization.

Satellite Spacing

With more than 20 C-band satellites now hovering over North America, the geostationary parking lot in the sky known as the Clarke Belt is becoming pretty congested. Some satellites are separated by no more than two longitudinal degrees. In order to prevent interference from such closely spaced C-band satellites, it is necessary that polarization formats alternate between adjacent satellites. If one satellite uses horizontal polarization for its even-numbered transponders and vertical polarization for its odd-numbered transponders, an adjacent satellite separated by only two degrees will have the opposite polarity format.

Receiving Systems

A home satellite receiving system, better known as TVRO for TeleVision Receive Only, consists of a parabolic dish, feedhorn, Low Noise Block (LNB) amplifier and a receiver.

The function of the dish is to collect microwave energy containing the original picture and sound (modulated) information from a targeted satellite and then reflect that energy to a focal point in front of the dish.

The feedhorn, which is located at the focal point, funnels the concentrated energy reflected by the dish to a probe (the actual antenna) inside it that can be rotated or electronically switched to pick up either vertically or horizontally polarized signals.

Microwave signals are coupled to the LNB through a waveguide on the feedhorn. The LNB is actually two components housed in one case. Its first responsibility is that of an amplifier to boost the very weak signals. Second, it serves as a downconverter to change all 12 transponders of whichever polarity is being received to a lower block of frequencies so that the signals can be sent to the receiver inside the house using inexpensive coaxial cable.

The satellite receiver in the house tunes the individual transponders by processing the downconverted block of frequencies and demodulating or extracting the original picture and sound information. The same images and sounds that originated in the studio are now available for direct connection to a television monitor and stereo system.

—— Satellite Basics Q & A ——

Ed. Note: Satellite TV is a complex and constantly changing field. One way dish owners keep up is through the author's "Ask the Tech Editor" column in Satellite TV Week. *Here is a sampling of letters that deals with changes in the last couple of years. They also help give a broader and more detailed understanding of satellite TV.*

Bird Spacing

Q: I'm a bit confused when it comes to satellite spacing. I thought that the minimum spacing between satellites in the Clarke Belt is three degrees with the current technology of receivers. Many satellites are closer than that, and now I've read that Galaxy 5 will replace Westar 5 at 125 degrees and Telstar 303 will shift to 123 degrees. That's only two degrees apart. The same situation repeats many times across the belt (F2 and G2, F1 and F5, etc.).

I've noticed too, that SBS-6 (Ku-band) is in the 99-degree slot which also is occupied by Galaxy 6 (C-band). Are there actually two separate birds in the same location or is it the same satellite with two different modes of transmission?

— Carter West, Texas

A: *The Federal Communications Commission established a two-degree spacing policy in 1983. The policy was initially opposed by several groups, including the National Cable Television Association*

(NCTA). At that time, cable TV antennas were designed for maximum gain and exhibited high sidelobes which would not permit the antennas to operate in a two-degree environment. The FCC gave the cable industry five years to either replace or retrofit its antennas to receive signals from satellites spaced at two degrees.

Since the policy was first established, it has been reaffirmed several times. The FCC recently denied two petitions for rulemaking to change its orbital spacing policies from two degrees to three degrees. The first paper filed was a petition from the Satellite Dealer Forum and K-Sat Broadcasting. The second was by a group of 12 entities that included satellite equipment manufacturers. Supporters argued that if satellite spacing were to remain at three degrees, the new generation of higher-powered (16-watt), C-band satellites would permit excellent reception of signals with a four-foot dish.

Unfortunately, four-foot antennas can't discriminate between two satellites parked two degrees apart. A four-foot antenna has a beamwidth of about 4.2 degrees. If this antenna were pointed at a satellite with adjacent birds separated by two degrees on either side, the dish would receive signals from all three satellites simultaneously and render the picture unviewable. Even six-foot antennas may experience problems when trying to target a satellite with another bird parked two degrees on either side. Although we will see an appreciable increase in power with the new generation of replacement satellites, I'm not recommending antennas smaller that 7½ feet in diameter.

For the majority of home satellite system owners with 10-foot antennas, two-degree spacing will not be a problem. However, if the dish is warped or the feedhorn is misaligned, unwanted sidelobes may pick up interference from an adjacent satellite. Satellite dealers and service technicians will need to take extra care to assure parabolic symmetry and proper feed positioning.

SBS-6 and Galaxy 6 are separate birds that are co-located at 99 degrees west. This is possible because SBS-6 operates at Ku-band frequencies and Galaxy 6 operates at C-band. G-Star 4 and Telstar 303 also share the same orbital location at 125 degrees west. Spacenet 1, 2 and 3 are hybrid birds with both C- and Ku-band transponders.

Orbital Locations

Q: I know our communication satellites are located in a geostationary orbit 22,300 miles from Earth, but where do these satellites reside in respect to Earth? Some say they orbit over the southwest United States. Others have said they orbit over the center (Colorado or Nebraska). Where do they orbit?

In regard to spacing and distance, how far apart do they float in space? Is it 500 feet or a half mile? What does two-degree spacing mean in relationship to distance from each other?

— Fred Schall, Virginia

A: *Geostationary communication satellites are parked at specific longitudinal positions directly above the equator. In other words, their latitude is essentially zero. If you were living on the equator, your dish would be pointing straight up. Here in the northern hemisphere, our antennas point south; they look out at the arc of satellites directly above the equator which provides services for North America. Longitudinal positions range from 69 degrees west (Spacenet 2) over the East Coast to 137 degrees west (Satcom C1) over the West Coast.*

You can look at a map or call your local airport to determine the longitude of your location. Then look at the positions of satellites arrayed in the centerspread of Satellite TV Week.

The satellite with the longitudinal position which is closest to your longitudinal location is the satellite closest to true south. When targeting that satellite, your dish will be at the zenith, or highest elevation point on the axis about which it rotates.

To understand the relationship between longitudinal spacing and distance between satellites, it helps to first understand a little about the relationship of longitude and distance here on the surface of the earth. The earth is about 24,000 miles around at the equator and it rotates on its axis once every 24 hours. That means if you were to step off the earth to an imaginary spot in space, you would watch about 1,000 miles of geography, or 15 degrees of longitude, pass by every hour.

Extending those longitudinal lines from the center of the earth out to a distance of 22,300 miles from the surface of the earth, nearly the distance around the earth, the spacing between each degree is about 459 miles. Satellites parked at two degree spacing are separated by 918 miles of space.

Winegard Mesh Dish
Courtesy of Winegard Satellite Systems

CHAPTER 2
Installation

The purpose of this chapter is not to give readers detailed information about how to install their own system, but to offer an overview of proper installation techniques that will help current dish owners to identify installation flaws before they cause problems. It also aims to provide prospective satellite system owners a sense of what to expect from a professional installation.

A properly installed satellite system can offer many years of entertainment with a minimum amount of periodic service. But professionalism and expertise varies among installers and technicians.

This was especially true in the earlier years of the industry when satellite dealers were first learning skills of the trade. Antennas were installed close to trees that within a few years grew and blocked signals. Inadequate concrete foundations became undermined by water, which caused antennas to shift. Cables buried directly in the ground instead of being sheathed in PVC soon fell victim to the sharp little teeth of underground varmints such as gophers and moles.

Today, professionals know how to do the job right the first time. The first step to installing a home satellite system is the site survey. This is where the dealer or technician will assist the home owner in choosing the best location for the dish, check the site for the presence of terrestrial interference and plan the entire installation.

Terrestrial interference (TI) results when a home satellite system receives unwanted microwave signals from a nearby land-based source operating in the same band of microwave frequencies as the satellites. The most common source of TI comes from microwave relays operated by the telephone companies, although airport navigation systems also can disrupt satellite reception.

Satellite dealers use either a spectrum analyzer or a portable test dish to detect possible interference at the proposed site. A more detailed explanation of the effects of terrestrial interference, as well as methods of avoidance and suppression, will be found in another chapter of this book.

The location for the dish must have an unobstructed view of the entire arc of satellites. Satellite signals will not go through buildings or trees. There are a number of survey tools used by installers to ensure a clear line of sight to every satellite.

A good satellite dealer will pay attention to the desires of the customer and explore all possible sites on the property so the customer can choose a location for the dish that will have the least aesthetic impact on the landscape. When real estate in the back yard is at a premium or when trees are in the way, a tall pole mount or roof mount may be in order.

The dealer must plan the entire installation before any work can begin. Soil conditions at the proposed antenna site have to be checked and cable routes need to be clearly marked for the installation crew. The location of all underground utilities or sprinkler systems must be determined and noted on the plans before any digging or trenching can be done. Consideration also should be given to any future plans for landscaping, gardens, pools or additions to the home.

A decision also will have to be made as to where the cable will enter the house, how many TVs will be connected to the system and the best cable route to each TV. Ideally, dish owners will want the capability of viewing any source, including the satellite receiver, VCR, cable or off-air antenna on every TV in the home. There are a number of accessories available which can be used to distribute sources and allow the user to control any component from anywhere in the house.

The most common type of antenna installation is a ground mount whereby an eight- to nine-foot pole measuring 3½ inches in diameter is planted in a hole which is filled with concrete. Although soil conditions will dictate the actual size of the hole and amount of concrete required, the average installation will require a hole at least three feet deep. It should extend a minimum of six inches beyond the frost line. A 16- to 18-inch-diameter hole is sufficient for most lightweight mesh antennas. Ten-foot antennas will require that five feet of pipe be exposed above the ground.

Before any concrete is poured, at least six feet of trench should be excavated next to the hole. This will allow a short section of PVC pipe and a 90-degree sweep elbow to be placed in the trench and aligned inside the hole so that it extends up alongside the pole. Otherwise the cable will be exposed and subject to damage from lawn mowers and weed-eaters.

Once the hole is dug, the pipe and PVC can be placed in the hole with a little backfill to prevent concrete from running out into the trench. The pole should have a small piece of scrap metal or rebar welded on the side towards the bottom to keep it from twisting loose in high winds. Wooden stakes or guy wires can be temporarily used to support the pole while concrete is poured into the hole around it. A carpenter's level should be used to check that the pole is perfectly vertical just after the concrete is poured and periodically while the concrete is setting.

The rest of the trench can be excavated and the cable installed in the PVC while the concrete is curing. All installations should include a grounding rod placed next to the dish a foot or so from the concrete base. A heavy wire should be connected between the grounding rod at the dish and the grounding rod at the house service box. This will ensure the ground potential at the dish and the house are equal and afford some protection against lightning-induced surges. A heavy copper strap is attached between the ground pole and the grounding rod next to the dish. The heavy ground wire between the dish and the house can be directly buried in the trench alongside the PVC. In addition, a grounding block should be used on the coaxial cable where it enters the house.

Depending on weather conditions and additives mixed with the concrete to expedite the curing process, it may be possible to complete the installation the same day. Often, the crew will finish the trenching, cabling and antenna assembly in the morning while the concrete is curing and be able to install the antenna on the mount, track the arc of satellites, program the receiver and instruct the customer on the operation of the system before the customer feels obligated to invite the crew in for supper.

Details about the outside components, weatherproofing, tracking adjustments and receivers will be covered in other chapters.

The information provided here should allow old as well as new dish owners to inspect certain aspects of the installation. Is there an unobstructed view of all satellites? Do or will trees block some of the signals? If so, is there a better location for the dish? What about the concrete foundation itself? Is the pole still vertical? Check it with a carpenter's level. Was the cable directly buried or sheathed in PVC? Did the installer ground the dish and the cable at the entry point into the house?

Spotting potential installation problems now could prevent expensive repairs later.

Installation Q & A

Ed. Note: Satellite TV is a complex and constantly changing field. One way dish owners keep up is through the author's "Ask the Tech Editor" column in Satellite TV Week. *Here is a sampling of letters that deals with changes in the last couple of years. They also help give a broader and more detailed understanding of satellite TV.*

Another Site Survey?

Q: As a dish owner I've used your publication for four years now and I wouldn't be without it.

Now I've got a problem since NBC and CBS have scrambled. I'm stuck in the mountains and can't get F2 or F1. (I get everything from F4 to G1.) I would gladly subscribe to network affiliates but I just can't get them and football is coming up.

Do you know of any timetables for F1 and/or F2 coming down, or relocating, or for the affiliates to shift to satellites I can receive?

I'd appreciate any information you have on this.

— S.R. Williams, California

A: *Satcom F1R is due to be replaced in the first quarter of 1993, but its replacement will be parked in the same orbital slot. I haven't seen a schedule for F2 or heard of any plans by the affiliates now using F1R and F2 to relocate to another satellite.*

Are you convinced that your dish is in the best location on your property for a clear line-of-sight to all of the satellites? If so, would raising the dish on a higher pole or tower help in receiving signals from those satellites at the ends of the belt? From your location in Redding, Calif., the elevation to F1R is over 40 degrees. You may want to call for another site survey and a second opinion as to the best location for the dish. Even if moving or elevating the dish will permit reception of K2, which is only one degree past F4, that would give you NBC on Ku-band.

Lightning Protection

Q: I live in an area where lightning storms often occur in the summer months. Can you give me a tip on how to ground a satellite system?

— M.E. Patrick, Alabama

A: *Grounding rods, straps and clamps should be available from any good hardware or building supply store. An eight-foot copper-clad iron rod will be sufficient for most applications, but sandy soil condi-*

tions may dictate a longer rod. Check with your local power company for advice.

Clean any paint and rust from around the base of your antenna ground pole where you can attach a clamp. Next, drive the rod into the ground alongside the concrete base and attach a heavy copper strap or wire between the ground pole and the ground rod with clamps. In some cases, the ground potential at the dish may be different than that at the house service box. I recommend you connect another heavy wire from the dish ground rod to the ground rod at the house service box. This will equalize any difference in ground potential between the two locations.

Since damaging currents can enter your system via the AC wiring as well as the coaxial cable and control wires, you also should install a lightning and surge protection device. There are several models available through satellite dealers which include a warranty to cover all your equipment.

Orbitron SST-8
Courtesy of Orbitron

CHAPTER 3
Antennas

At the heart of the satellite system is the dish or parabolic reflector. Its purpose is to gather very weak signals from a targeted satellite and focus those signals to a precise point in front of the dish while ignoring signals from adjacently-parked satellites. Performance is directly proportional to the dish's size, parabolic symmetry and surface accuracy.

Four-section mesh antennas are the most popular because they are relatively inexpensive, have the least aesthetic impact and are easy and quick to assemble. Hydroformed, spun aluminum and sectional fiberglass antennas generally offer a smoother, more accurate surface and are more efficient at the higher Ku-band frequencies.

A typical mesh antenna is made up of four sections that come preassembled with the mesh already attached. Each section will feature four or five pre-formed ribs. The better antennas will have one or more intermediate rings fixed between each rib for additional support. Some manufacturers also offer models with eight sections or other designs that allow for shipping via UPS. Regardless of the design, proper assembly is a must to achieve top performance with any dish.

Most four-section mesh antennas are designed to be assembled on a flat surface. Two quadrants placed face down are first joined together with the hardware just finger tight. The procedure is repeated for the other two quadrants and then the two halves are joined together. The tightening sequence usually starts at the outer perimeter and moves row by row toward the center. Care must be taken to ensure that the seams between the adjoining outside ribs of each quadrant are flush. About 10 foot pounds of torque on the bolts is sufficient; the object is to stop just short of deforming or crushing the ribs.

Once the antenna quadrants are assembled, the structure should be turned over and checked for parabolic accuracy. The simplest way is to sight across one lip to the lip on the opposite side. They should appear perfectly parallel. If one edge of the lip is higher or lower than the other, the dish is out of shape and corrective measures must be taken.

A more accurate method is to check the dish by stretching two strings tightly across the lip-to-lip surface at right angles to each other so they intersect in the center. The strings should just touch where they intersect. If the strings are mashed together or separated by a gap, it is an indication that one edge of the lip is higher or lower than the other and the dish is not perfectly parabolic.

This test can be performed while the dish is on the mount as well. It may even be possible to loosen the hardware between the sections, adjust the panels to align the seams and retighten the hardware without having to take the dish off the mount. A clamshell or warped dish is often caused by improper assembly procedures but also can result from heavy winds and snow loads.

Another factor equally important to antenna performance is the smoothness of the surface. Satellite signals collected by the parabolic surface of the dish are reflected to a precise focal point in front of the dish where they are concentrated. Any irregularities in the reflective surface, such as dents or bubbles in the mesh panels, will cause signals reflected from that section to miss the focal point and reduce the antenna's efficiency.

A dish that is warped or has bent panels can suffer from more than just loss of performance. Irregularities in the parabolic surface can pick up signals from an adjacently-parked satellite and reflect those signals to the focal point as well. With satellites now parked as close as two degrees apart, the ability of a dish to focus on one satellite while rejecting signals from an adjacent satellite has become very important. Many mesh antenna manufacturers have incorporated preformed mesh to more accurately conform to their design curve.

One design aspect that should be understood by system owners is the antenna's focal length to diameter ratio (f/D). The focal length is simply the distance measured from the bottom of the parabolic curve out to where the microwaves are concentrated at the focal point. Where the focal point is located depends on the diameter of the dish and the depth of the curve.

Some antennas are designed with a relatively shallow curve and others are deep. Shallow dishes generally have a little more gain or signal gathering ability than deep dishes of the same diameter, but deeper dishes have more ability to reject terrestrial interference. Since precise location of the feedhorn is necessary for optimum performance, it's a good idea to know how to calculate the focal distance and f/D for your dish.

The formula for focal length is "f equals diameter squared, divided by 16 times the depth." The depth of the curve is measured by placing a straightedge across the lip-to-lip surface and measuring from the center to the bottom of the curve. Be aware that on some antennas the bottom of the parabolic curve may actually be cut off by a center plate or hub, so you may have to estimate.

Calculating Focal Distance and f/D Ratio

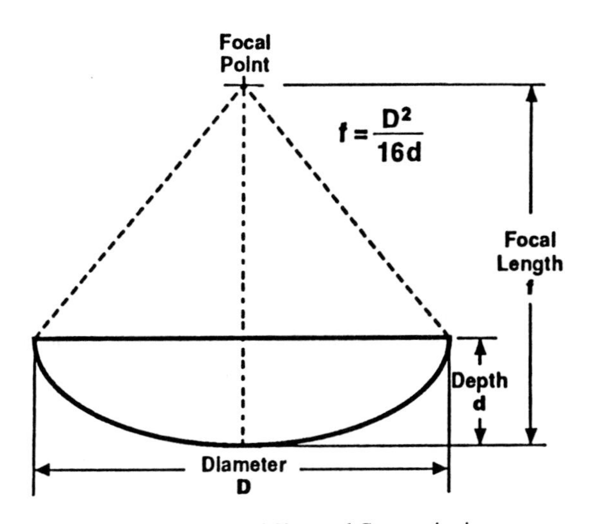

Diagram courtesy of Chaparral Communications

As an example, let's say the dish is 10 feet in diameter and the depth is 20 inches. First, we convert the 10 feet to inches and then square that figure. That's 120 squared or 14,400. Then we take the depth (20 inches) and multiply it by 16. That's 320. The formula should now read "diameter squared (14,400) divided by 320." The result is 45. The focal distance from the bottom of the curve to the edge of the feedhorn located at the focal point should measure 45 inches.

So what is the f/D of this dish? The f/D is the focal distance divided by the diameter, or 45 divided by 120 which equals .375.

If the f/D ratio and dish diameter were already known, the focal distance could be easily calculated without having to measure the depth of the antenna. The focal distance would simply be the f/D ratio

(.375) times the diameter in inches (120) which equals 45 inches.

Knowing how to determine the focal distance and f/D ratio permits a dish owner to check a dish to make sure the feedhorn is located at the correct focal point. This information will be used again in learning the importance of matching the feedhorn to the dish in the next chapter.

—— Antennas Q & A ——

Ed. Note: Satellite TV is a complex and constantly changing field. One way dish owners keep up is through the author's ''Ask the Tech Editor'' column in Satellite TV Week. *Here is a sampling of letters that deals with changes in the last couple of years. They also help give a broader and more detailed understanding of satellite TV.*

Cool Dishes

Q: I have an 11-foot metal dish which was painted a light yellow when I purchased it. In order to make it blend in with the house, I painted in a dark brown.

My TV man says the dark color is reflecting heat up into the pickup receiver, and that I should paint it a light color. This doesn't sound right to me, as I've always thought the light colors reflected more heat. What's the catch?

— L.H. Applegate, California

A: *You are correct in thinking that a light-colored dish will reflect more heat than a dark-colored dish. With mesh antennas the solar energy that is reflected will rarely cause a problem with the electronics at the feedhorn. Solid dishes are another story.*

A solid dish can actually act like a solar furnace, especially if it is a shiny, bright color. The most dangerous times of the year are during the equinox in the spring and fall when the sun passes across the equator. On those days all satellite downlinks experience solar outages for several minutes as the sun passes behind each satellite. On the worst day the outage may last for 10 minutes and if you look at your dish you will notice that the sun is casting a perfect shadow of the feed in the center of the dish. For those of you with shiny, solid dishes, you can avoid possible damage to the electronics located on the feed by pointing your dish at western satellites in the morning and eastern satellites in the afternoon.

G5 Bombs T3

Q: Ever since G5 was activated, I've had reception problems with

T3. What's going on? I'm using an eight-foot dish.

— Mike Reynolds, Oklahoma

A: *Telstar 3 was relocated from 125 degrees west to 123 degrees west. Galaxy 5 now is parked in T3's old location at 125 degrees west. This bird dance was necessary to satisfy the FCC's two-degree spacing requirements. By now, everyone should have these changes programmed into their receivers. If not, you'll be receiving G5 in the wrong polarity format when you recall T3.*

Every satellite dish has a beamwidth, usually measured between half-power (3dB) points, which determines the acceptance angle of the antenna or the amount of sky it "sees" when targeting a specific satellite. Typically, a 10-foot antenna will have a beamwidth of about 1.6 degrees. A good 1½-foot antenna will have a beamwidth of 2.4 degrees. That is, if the parabolic shape is accurate and there are no bubbles or dents in the surface that will effect the sidelobes or off-axis response of the antenna.

Even with a 10-foot antenna, if the dish is warped or the feedhorn is not centered and looking squarely into the center of the reflector, its ability to reject signals from an adjacently parked satellite will be hampered. The chance of receiving co-satellite interference is even greater when you are targeting a satellite like T3, which has 8.5-watt transponders, and signals from G5 with its 17-watt transponders fall within the main beamwidth of the dish. Many six-foot dish owners have been experiencing problems with reception of signals from F3 (F1R) ever since G1 was moved to 133 degrees west.

Trouble With Ku-Band

Q: We have a GI 2730R system that works great on C-band. However, Ku-band is always a problem. We can get K2, but the reception is not that great and constantly has to be retuned. Every now and then we find something on the other Ku-band satellites, but it disappears the next day. Any hints?

— David L. Shafer, Missouri

A: *Antenna symmetry, feed positioning and antenna tracking are three times more critical for Ku-band reception than for C-band. If the antenna is warped or has any irregularities in the surface such as bent or bubbled panels, Ku-band performance will be affected. The feedhorn must be perfectly centered, set to the proper focal distance and looking squarely into the center of the dish. Guy wires are recommended to stabilize the feed on antennas with buttonhook-type feed supports. In addition, excessive play in the mount axis or motor drive* *will not permit the antenna to accurately return to preprogrammed positions.*

You can check the parabolic shape of the reflector by sighting across the lip-to-lip surface of the dish. It there's any obvious bow or twist in the edges, it will have to be corrected. A more accurate method for checking the apabolic symmetry consists of placing strings across the edges of the reflector from top to bottom and side to side. The strings should just touch where they intersect in the center.

Feedhorn centering can be checked by measuring from the edge of the feedhorn to the lip of the dish from three different locations 120 degrees apart. All measurements should be equal. An inclinometer or Arc-Set tool can be used to verify that the scalar plate is parallel to the lip-to-lip surface of the dish and looking squarely into the center.

Antenna tracking also becomes three times more critical for reception of Ku-band signals. Tracking consists of making adjustments to the axis elevation declination and azimuth. Competent installers who have experience with Ku-band use a very sensitive signal strength meter to aid in precisely aligning the mount angles and azimuth heading so that the reflector can track all satellites in the arc with Ku-band accuracy.

You may be able to straighten out any bent panels, center the feedhorn and tighten loose hardware, but I recommend you contact a dealer with Ku-band experience to fine tune the feedhorn and tracking adjustments.

Ku-Band Query

Q: I would like to update my system for Ku-band, but do not know if my system will be compatible.

My antenna is a Continental C-105, 10½-foot, 3.2-meter antenna. Gain is 4.0 GHz .85 degrees at -3dB. The material is aluminum mesh. It has eight inner screens (points) and eight outer screens (bases). I measured the diamond-shaped holes in the mesh (same for inner and outer screen sections) as ¼-inch wide and 1/16-inch high. My receiver is a Luxor 9900, which I am very happy with. I receive excellent pictures on C-band and get great audio reception.

My questions are as follows:

Will the antenna work satisfactorily on Ku-band?

If not, are replacement panels available?

Will I have to replace both inner and outer screens?

Will my Luxor 9900 offer satisfactory Ku-band reception?

Will my Thompson Saginaw 18-inch actuator give satisfactory Ku-band service?

What degree Ku-band LNB would you recommend using dual feed?

Any preference on the C/Ku-band feed? I realize that there is a limited amount of programming on Ku-band; but in Fresno, what satellites can I expect to receive from your magazine's "Ku-Band Channel Choice" chart? Anik C-1? K2? One of my main pleasures on C-band is the audio hidden on the transponders. Are such audio channels also available on the Ku-band, both regular and SCPC? (Incidentally, I think your answer would be of interest to many of your readers.)

— John Oliveira, California

A: *Your Continental C-105 antenna will need a little work to make it Ku-band compatible. The aluminum mesh size is fine, but the antenna efficiency is only about 20 percent at Ku-band because the inner screens did not conform to the parabolic symmetry of the dish. This was primarily due to the elimination of a support ring in an effort to cut production costs. The problem can be corrected by rolling out the inner screens so that they follow the curve of the ribs and securing the screens to the ribs. Dan Berge at TelSat International, 503-656-2774, can give you some tips on how to go about forming the inner screens.*

With a little keystroking to get the synthesizer into the correct channel, your Luxor 9900 should work well on Ku-band. A few of the early units experienced some C-band interference on unused Ku-band channels as a result of a problem with the internal switch, but it is correctable. Actuators that only deliver 20 or so pulses per inch of travel may not accurately return the dish to preprogrammed Ku-band positions. Once the accuracy of the dish is corrected, a 1.6dB or 1.8dB LNB should do fine. As for feeds, I like the Chaparral Corotor of the California Amplifier Centerline. You should be able to pick up all the Ku-band birds with the exception of the Canadian satellites. And, yes, you can find SCPC signals on some of the Ku-band satellites.

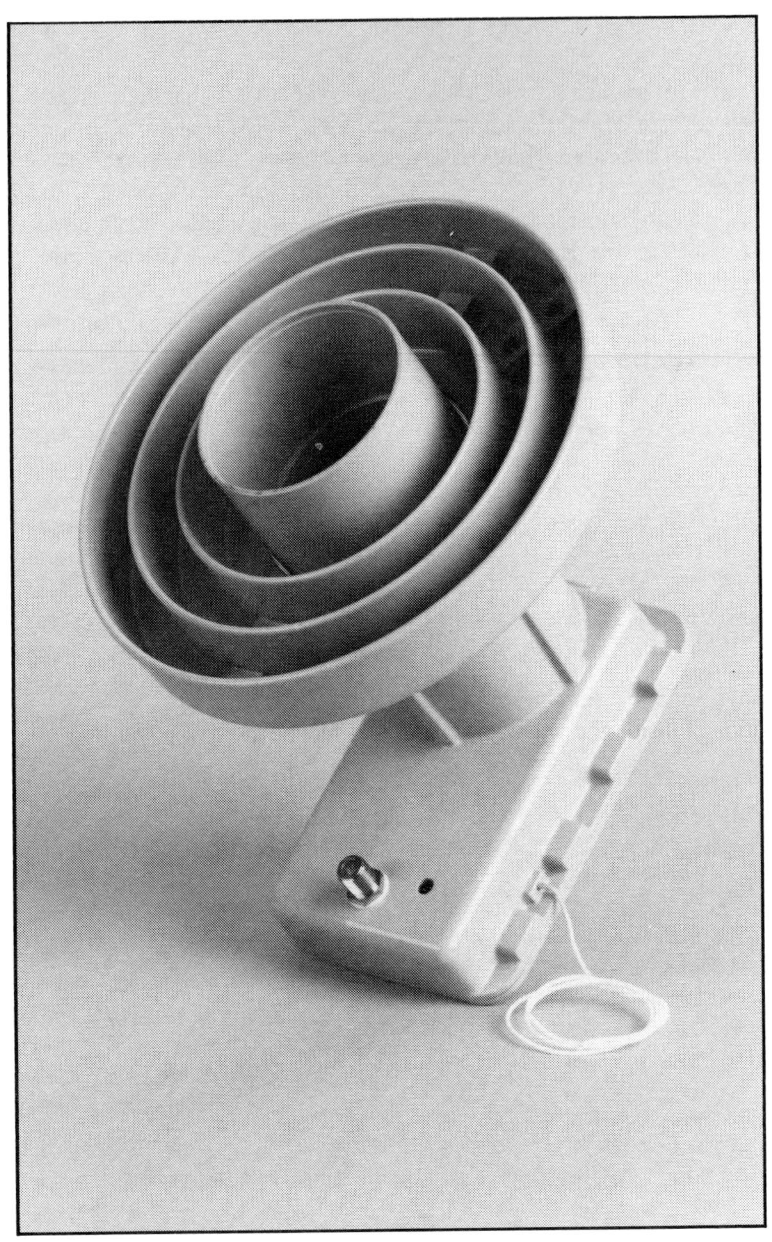

California Amplifier C-band LNBF
Photo courtesy of California Amplifier

CHAPTER 4
Feedhorns

The feedhorn located at the focal point in front of a prime focus satellite antenna has the responsibility of collecting microwaves reflected from the surface of the dish while rejecting noise and other signals coming from off-axis directions.

Energy concentrated at the focal point is directed via a circular waveguide to a small probe (which is the actual antenna) precisely positioned inside the waveguide. In most feedhorns, the position of the probe is controlled by a small servo motor to determine which sense of polarity (vertical or horizontal) is detected. The signal is then funneled through a waveguide elbow and coupled to the LNB which amplifies the very weak signals and downconverts them to a lower block of frequencies which are tuned by the satellite receiver or IRD located in the house.

C-band feedhorns also have a scalar plate made up of concentric rings that surround the circular throat or waveguide. The position of the scalar around the circular waveguide determines the feedhorn's field of view or how it "reads" the dish. In order to properly illuminate the dish, the feedhorn must be optimized to match the f/D ratio of the dish.

In the previous chapter on antennas we learned how to calculate the focal point of a dish given the diameter and depth of the dish and how to determine the focal length to diameter (f/D) ratio. Deep dishes have a focal length to diameter ratio from .027 to .032 and a wide illumination pattern. Shallow dishes with an f/D ratio ranging from 0.33 to 0.45 have a narrow illumination pattern and their focal points are further from the dish.

Feedhorns are available with either fixed or adjustable scalars.

Most feedhorns with fixed scalars are optimized for dishes with an f/D ratio of .375. Feedhorns with adjustable scalars can be adjusted to precisely match the f/D ratio of any dish by sliding the circular waveguide in and out of the scalar.

It's important that the feedhorn is located at the focal point and looking squarely into the center of the dish. Otherwise, some of the signals reflected by the surface of the dish will miss the feedhorn. If the feedhorn is located too close to the dish, it will be under-illuminated and will not be able to gather all the energy reflected by the dish. If it is located too far away from the dish, the feedhorn also will pick up noise from the surrounding ground. In either case, performance will suffer.

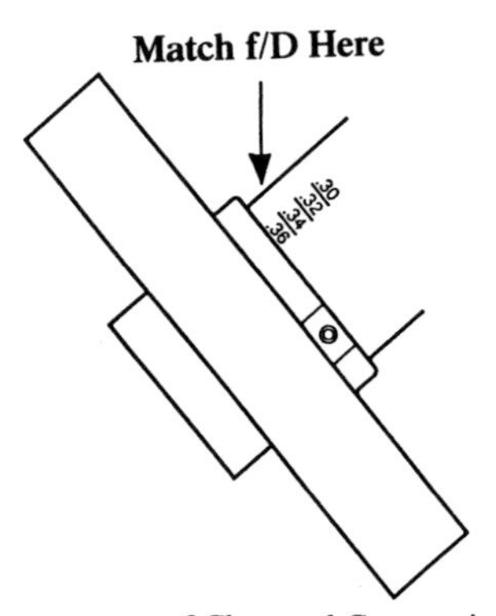

Diagram courtesy of Chaparral Communications

The first step to optimizing the feedhorn is to adjust the scalar to match the f/D of the dish. The feedhorn also must be oriented with the polar axis so that the probe will be able to rotate to select either polarity on any satellite without running into the mechanical stops. Manufacturers provide a template for aligning the feedhorn. Adjustable feeds will have either locking nuts or a threaded scalar and an f/D scale on the throat.

The next step is to set the focal distance measured from the bottom of the dish's curve to the tip of the circular waveguide throat. If the bottom of the curve is cut off by a center plate or hub, use the manufacturer's suggested measurement from the center plate to the end of the circular waveguide.

Confirm that the feedhorn is located at the center by measuring from the edge of the feedhorn to three locations around the perimeter of the dish at 120-degree intervals. All three measurements should be the same. You also can use an angle measuring tool such as an inclinometer to make sure the angle of the feedhorn throat and scalar match the angle of the dish. If you can't find a surface on the back of the dish you can trust to be parallel to the lip-to-lip surface of the dish, place a long straightedge across the lip-to-lip surface or a quadrant. Take your measurement down the steepest slope and then transfer the tool to the feedhorn and see if it's the same angle.

Scalar Ring Adjustment and Illumination

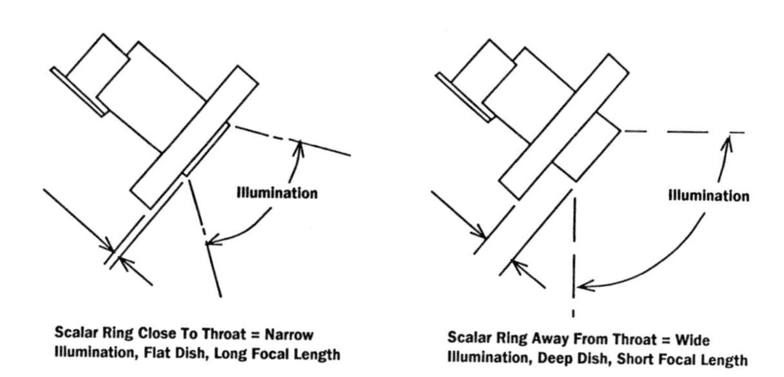

Diagram courtesy of Chaparral Communications

There are a number of sources for specialty tools, which include Arc-Sets, Focal Finders, signal strength meters and other instruments for fine tuning satellite dishes and feedhorns. Tool sources are listed in the back of this book.

The most popular feedhorn for reception of C-band signals is the Chaparral Polarotor 1, although there are similar models manufactured by Astrotel, California Amplifier, Fujitsu General and others which use a small servo motor to select polarity.

All popular receivers provide drive circuits to control mechanical feeds with servo motors. One terminal on the rear of the receiver supplies five volts for powering the servo, another is for the pulse which changes the sense of polarity and a third terminal is ground.

California Amplifier and Chaparral Communications also offer dual-band feedhorns for the reception of both C- and Ku-band signals. Cal-Amp has two C/Ku-band feedhorns. One is designed for deep dishes and the other is optimized for dishes with a 0.375 f/D ratio. The Chaparral Corotor II Plus C/Ku-band feed features an adjustable scalar.

The Chaparral Polarotor 1 and the Corotor II Plus come with a 90-degree waveguide elbow which bolts to the waveguide on the feedhorn. This configuration allows the LNB to be bolted to the other end of the elbow to maintain a slim profile and fit inside the protective plastic feedhorn covers supplied by antenna manufacturers. Gaskets must be used between the feedhorn waveguide and 90-degree elbow, and between the 90-degree elbow and the LNB. Otherwise, water can enter the elbow and block the signal. Worse yet, water that enters the waveguide can freeze. When water freezes it expands and may crack the waveguide.

Feedhorn - Cross Section

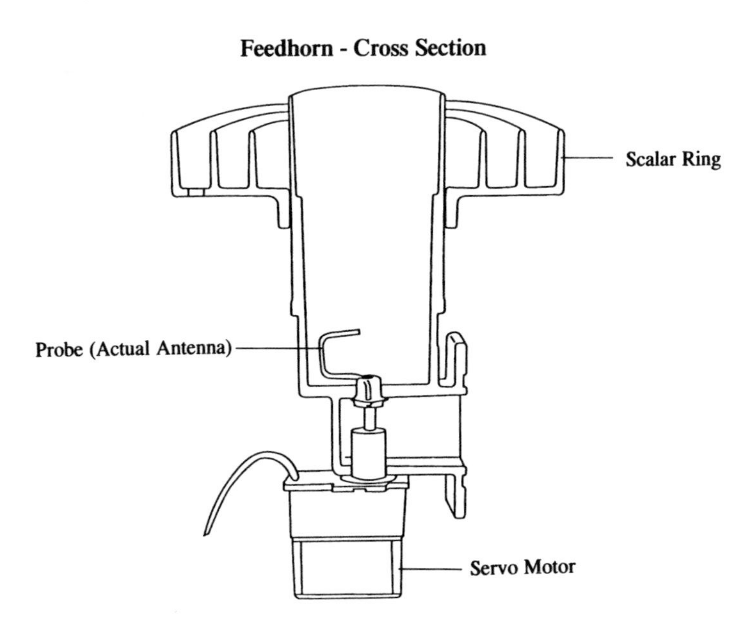

Another place water can enter the feedhorn is via the seal around the servo motor. This was a problem when servo-operated feedhorns were first introduced. Manufacturers had to include a plastic protective cover for the servo to prevent water from seeping into the waveguide past the servo seal. Since those early days seals have improved. Virtually all antennas are now shipped with a feedhorn weather cover that fits over the entire feed assembly.

All feedhorns include a plastic cap to cover the throat of the circular waveguide that looks down into the dish. The plastic cap does not interfere with reception because microwaves go through plastic. However, without the plastic cap, wasps and other insects can build their homes inside the throat of the feed. That will block the signal. Be careful not to touch the probe when removing foreign objects from inside the throat. As mentioned before, the probe is the actual antenna and its position is critical to reception.

Common problems related to feedhorns include misalignment, water intrusion, blockage inside the throat from wasp or small bird nests, servo motor failure and bad servo wire connections.

It's a good idea to check the condition of the gaskets between the feedhorn waveguide and 90-degree elbow, and the elbow and the LNB. If they are deteriorated or hardened, replace them. To prevent water from accumulating in the waveguide elbow, you can drill a very small hole in the lower corner of the elbow for drainage.

The plastic cap covering the throat of the feed should have several small holes to allow moisture to evaporate. Remember, water blocks microwaves. If your feedhorn does not have a plastic cover that protects the entire feed assembly from snow and rain, get one and install it. They're not expensive and are available from your local satellite dealer.

Several innovative feedhorn designs have appeared recently. Some manufacturers incorporate the 90-degree elbow in the feedhorn casting to reduce assembly time and improve coupling between the feedhorn and elbow. Other manufacturers ship the feedhorn fully assembled with a matched LNB.

Until a year or so ago, there was at least one satellite that broadcast its signals with the polarity skewed at about a 22-degree angle compared to other birds. This made it necessary to have a way of adjusting the polarity probe beyond the usual 90 degrees to accommodate both vertically and horizontally polarized signals.

That is no longer the case. With a polar mount antenna and the feedhorn properly orientated to the polar axis, skew is not really necessary. As the antenna rotates about its polar axis to look at satellites to the east or west, skew becomes automatic.

This has led to the introduction of C-band feedhorns without a servo motor to move a probe. Instead, the new feedhorns feature two pickups, one for vertical and one for horizontal, and switching between polarities is electronic with no moving parts. Some of these new feeds use the same pulse wire that operates the servo to change polarity. Others are designed to switch polarity with a 13/18 volt change on the LNB cable.

—— Feedhorns Q & A ——

Ed. Note: Satellite TV is a complex and constantly changing field. One way dish owners keep up is through the author's "Ask the Tech Editor" column in Satellite TV Week. *Here is a sampling of letters that deal with changes in the last couple of years. They also help give a broader and more detailed understanding of satellite TV.*

Weak Ku-Band Signals

Q: I recently purchased an STS SR-150 IRD, SAMI 10-foot dish, Chaparral Corotor II feedhorn, 35-degree C-band LNB and a 1.2 dB Ku-band LNB.

It seems as if I miss a lot of transponders because the signals appear to be too weak and I have only been able to pick up one Ku-band satellite (K2).

Is this normal or could this be a problem with the installation and setup? In reading the documentation for the feedhorn there was mention of f/D ratio and centering the feedhorn. I watched the installation and I don't recall such things being done. How important are these steps? What do you recommend I do to improve my system performance?

— Tyrone M. Stanton, California

A: *Dish assembly, feed positioning and antenna tracking are critical for Ku-band reception. First of all, the dish must be correctly assembled so that it is parabolic. All seams between the sections must be flush and there should be no irregularities or bubbles in the mesh. A parabolic dish is designed to reflect microwaves from a targeted satellite to the focal point in front of the dish. Microwaves reflected from a section of the dish that is not parabolic will miss the feedhorn, resulting in reduced efficiency.*

Likewise, the feedhorn must be located at the proper focal distance and look squarely into the center of the dish. If the feedhorn is not precisely aligned at the focal point, it will not be able to gather all the energy reflected by the dish surface. Again, efficiency will be reduced. Your Corotor II feedhorn has an adjustable scalar that should be set to match the f/D ratio of your dish.

Antenna tracking consists of adjustments to the declination, elevation and azimuth. Professional installers use an inclinometer or Arc-Set tools and a sensitive signal strength meter to optimize the feedhorn and mount adjustments that are particularly critical to good Ku-band reception.

I suggest you contact your dealer and let him know about your dissatisfaction. The equipment you purchased has the capabilty of producing excellent reception from both C- and Ku-band satellites.

Feedhorn Centering Tool

Q: In the July 28 to August 3 edition of *Satellite TV Week*, you reported in "Wavelengths" about a company that manufactures two feedhorn-centering devices. The first device was a telescoping rod and the second was a laser beam. My copy of the edition was discarded prior to my writing down the name, address and phone number of the company. Could you please research this information and send it to me?

— Eugene Gunn, Florida

A: *Feedhorn centering is crucial to system performance and should be checked periodically, especially after heavy snow or wind storms. Energy reflected by the dish is concentrated at the focal point in front of the dish. If the feedhorn is not centered and located at the proper focal distance, some of the reflected energy will miss the feedhorn and performance will be degraded.*

The tool you are looking for is the Focal Finder from Natropolis International, P.O. Box 14115, St. Paul, Minn. 55114, 612-646-4700.

Probe Positioning

Q: I am getting signals on only one polarity on my VideoCipher II 2400R receiver. After replacing the polarizer motor with a new one, I ended up getting two signals on one channel or in-between signals. The slot was lined up with the polarized motor. What am I doing wrong?

— S. Eller, Washington

A: *The small servo motor on the back of your feedhorn controls the position of the probe (actual antenna) so that either polarization can be detected. When switching from an even to an odd channel, or an odd to an even channel, the probe rotates 90 degrees. The skew control on your receiver is used to compensate for signals that arrive off axis and will allow the probe to travel back and forth through about 140 degrees of movement.*

Probe movement is determined by the motor control circuit and width of the pulses. The motor has mechanical stops and can move only so far. If the pulses controlling the probe are too short or too long, the probe will run up against the mechanical stops and eventually burn out the servo motor.

Since the total probe movement is restricted to 140 degrees, the feedhorn must be oriented with the polar axis of the dish so that both polarities can be tuned without having the servo motor run up against its stops. Chaparral provides a template with its feeds that slips over the servo motor and has an arrow that points to the polar axis. You should be able to get one from your local dealer or by calling Chaparral at 408-435-1530. Unless you changed the probe position when installing the new servo motor, all you should need to do is slip the template over the motor and rotate the feed until the arrow points to the polar axis of the dish. If a problem still exists, have your dealer check the receiver's polarity drive circuit.

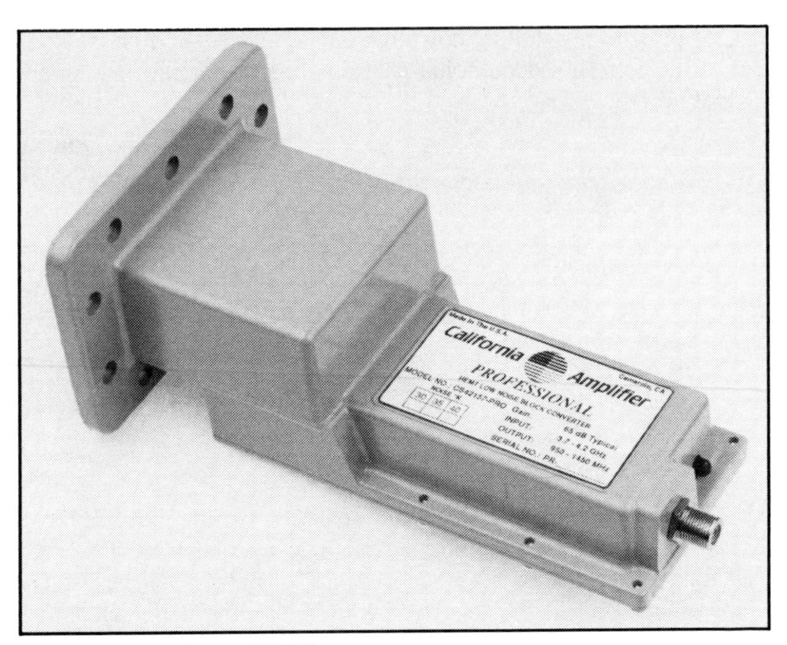

C-band LNB - Professional Model
Photo courtesy of California Amplifier

CHAPTER 5
LNBs

Low Noise Block downconverters (LNBs) attach to the waveguide or waveguide elbow on the feedhorn. The purpose of the LNB is to amplify the very weak signals reflected by the dish and passed through the feed, and to downconvert the microwave signals to a lower block of frequencies which can be sent to the receiver in the house for processing. LNBs amplify the signals by a factor of 100,000 (50 dB) or more and are rated in Kelvin degrees, which is an indication of how much noise (static or video ''snow'') is introduced by the amplifier. The lower the degree, the better the performance.

First-generation home satellite systems used a Low Noise Amplifier (LNA) and a separate downconverter. Typically, these systems used a single downconversion scheme whereby the output from the downconverter contained a single channel of information. Manufacturers of 70 MHz, single downconversion receivers used various designs, so downconverters were not standardized or interchangeable.

Eventually, the industry developed a standard downconversion scheme which is a block system with an output which ranges from 950-1450 MHz. Virtually all U.S. receiver and IRD manufacturers have adopted this downconversion scheme.

With a block downconverter, all 12 vertical transponders or all 12 horizontal transponders (depending on the polarity selected with the feedhorn probe) are downconverted as an entire block of frequencies.

Since the output of a block downconverter contains the channel information for all 12 transponders of whichever polarity is selected, it's possible to have multiple receivers connected to the same block downconverter or LNB, each tuning a different channel or transponder. It's also possible to install a dual-polarity feed with one LNB processing vertical signals and the other LNB processing horizontal signals. Using a multi-receiver switch, each receiver can independently tune any vertical or horizontal transponder on the same satellite. This is how a local cable company picks up multiple channels off the same satellite with only one dish.

The adoption of the 950-1450 MHz block downconversion scheme by the industry led to the production of the LNBs used today. An LNB is simply a Low Noise Amplifier (LNA), like those used in conjunction with early downconverters, combined with a block downconverter in the same housing. Advancements in electronics, particularly the development of the High Electron Mobility Transistor (HEMT), have allowed LNB manufacturers to produce much more efficient and durable products. Typical noise ratings on today's LNBs range from 25 degrees Kelvin to 45 degrees Kelvin, whereas the industry was using 100 and 120-degree LNAs in the early 1980s.

A few manufacturers have introduced Low Noise Block Feeds (LNBFs). These products combine the LNB and feedhorn in one package for simpler installation. Improved performance is claimed because of better matching between integrated components.

Water intrusion in the cable connector and lightning are the most common causes of LNB failures. The coax cable connector should be packed with a dielectric sealant or sealed with Coax Seal to keep water from seeping into the connector. There are protection devices designed to suppress lightning-induced surges, which we will cover in another chapter.

—— LNBs Q & A ——

Ed. Note: Satellite TV is a complex and constantly changing field. One way dish owners keep up is through the author's "Ask the Tech Editor" column in Satellite TV Week. *Here is a sampling of letters that deal with changes in the last couple of years. They also help give a broader and more detailed understanding of satellite TV.*

Half the Channels

Q: I have a General Instrument 2600 R receiver. Subsequent to moving it from western New Mexico to north central New Mexico, I have been unable to receive channels 1 through 10 on any C-band satellite. Channels 11 through 24 work perfectly and Ku-band works correctly on all channels.

GI says the problem is a defective LNB, but I would have thought that a defective LNB would cause deterioration of reception across the spectrum of bands and channels.

LNBs are not inexpensive, so I would like to get a second opinion prior to replacing it.

— Guy A. Mathews, New Mexico

A: *The symptoms you describe do suggest a faulty C-band LNB. However, a bad coaxial cable or C/Ku switch also could cause the attenuation of the lower channels.*

Since you have two coaxial cables, one for the C-band LNB and one for the Ku-band LNB, try switching the cables both at the LNBs and at the C/Ku switch. Just make sure you unplug the power cord from the AC receptacle before swapping the cables because there is voltage applied to the LNBs when the receiver is in the standby mode. Inspect the cable connectors when you take them loose. Look for any evidence of water intrusion or improperly installed connectors. If you're able to tune all the C-band channels and not the Ku-band channels, you've got a bad cable or connector. Otherwise, suspect a faulty C-band LNB or C/Ku switch.

The easiest way to confirm a bad LNB or C/Ku switch is substitution. Most satellite dealers stock both items. You could either schedule a service call or purchase an LNB and install it yourself. If you choose the latter, ask the dealer if he will give you credit toward a service call in the event the LNB doesn't fix the problem.

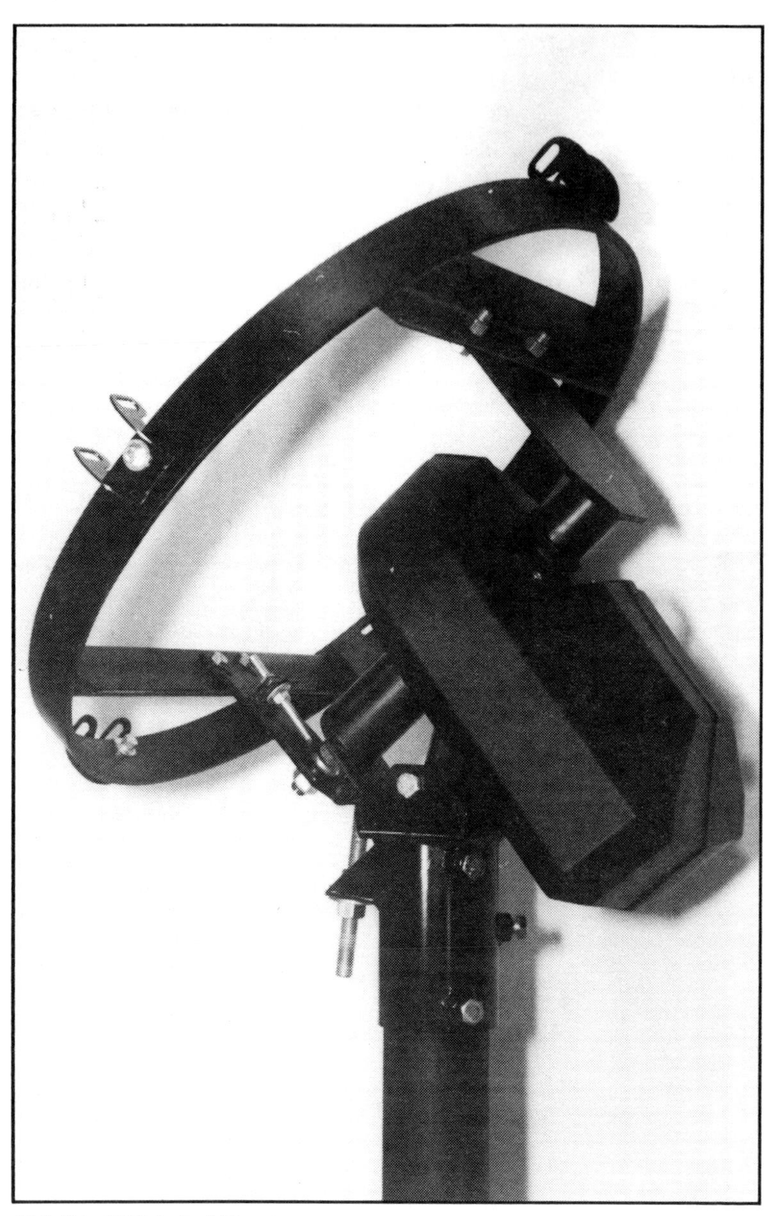

Ajak H to H Motorized Mount
Photo courtesy of Ajak Industries

CHAPTER 6
Mounts

Polar Mounts

The majority of movable home satellite dishes have a polar mount, though some have a horizon-to-horizon mount. The primary function of either mount is to accurately aim the antenna toward any satellite in the arc. The mount also provides support for the antenna and stability when targeting a chosen satellite.

Typically, polar mounts used for mesh antennas consist of a mount cap that fits over the ground pole, a polar axis assembly, elevation and declination adjustment hardware and an antenna mounting ring.

The most common cause of weak signals or ''sparklies'' that appear in the picture is the inability of the mount to accurately point the antenna at a satellite. Three tracking adjustments are necessary to match the mount rotation arc with the arc of satellites. These adjustments include the elevation of the axis about which the antenna rotates, the offset declination or amount the antenna is tilted forward from the axis elevation and the true south heading of the axis.

If a dish is moved up to its highest elevation where it is squarely positioned over the mount (balance point), it can be seen that the angle of the dish is tilted slightly forward from the angle of the axis. This declination offset is necessary because the arc of satellites is actually flattened at the top like an ellipse.

There are a number of tools available which are used by professional installers for setting the critical mount adjustments necessary to achieve accurate tracking. Some technicians use a compass and inclinometer and will set up a TV at the dish to monitor their results. Others are accustomed to using Arc-Set tools and a sensitive signal strength meter which responds to very small changes in signal level long before a viewer would be able to detect a change in the picture on a TV screen.

The Arc-Set has three angle duplicating tools with bubble levels much like a carpenter's level. One tool is set for the axis elevation, one for the declination and one for the elevation of the lowest satellite in the arc. The tools are adjustable and can be ordered preset for any location or adjusted by the user to match the mount angles of a properly tracked antenna in the same local area.

If a dish installer is going to use an inclinometer for setting the mount angles, he or she will need to know the latitude of the location and select the elevation angles for the axis and declination in the chart below.

Polar Mount Alignment Values

LAT.	AXIS	ZENITH	LAT.	AXIS	ZENITH	LAT.	AXIS	ZENITH	LAT.	AXIS	ZENITH
80	80.22	88.69	50	50.69	57.31	40	40.71	46.27	30	30.63	34.96
75	75.33	83.64	49	49.70	56.21	39	39.70	45.15	29	29.62	33.83
70	70.43	78.52	48	48.70	55.12	38	38.70	44.03	28	28.61	32.68
65	65.52	73.32	47	47.70	54.02	37	37.69	42.90	27	27.59	31.53
60	60.59	68.06	46	46.71	53.92	36	36.69	41.78	26	26.58	30.38
55	55.66	62.72	45	45.71	51.82	35	35.68	40.65	25	25.57	29.23
54	54.67	61.64	44	44.72	50.72	34	34.67	39.52	20	20.47	23.45
53	53.67	60.56	43	43.72	49.61	33	33.67	38.38	15	15.37	17.62
52	52.68	59.48	42	42.72	48.50	32	32.66	37.25	10	10.26	11.77
51	51.69	58.39	41	41.71	47.38	31	31.64	36.11	5	5.13	5.89

@ 1986 Gourmet ... ENTERTAINING (213) 666-2728 [DECLINATION] = [ZENITH]-[AXIS]

The first step is to actuate the dish up to the highest elevation where it is centered over the mount and looking at true south. Place the inclinometer or Arc-Set tool (marked Axis or No.1) along the mount's polar axis line and adjust the elevation with the elevation rod until the inclinometer indicates the correct axis angle (measured from the horizon) or the bubble in the Arc-Set Axis tool is centered. It may be necessary to loosen hardware on the axis assembly to make an elevation change.

Next, place the Arc-Set tool marked Zenith or No. 2 on a surface of the dish, such as the center plate, which is parallel to the lip-to-lip surface of the dish. Then adjust the declination until the bubble is centered in the Arc-Set Zenith tool. Spun aluminum and hydroformed antennas may require that a straightedge be placed across the lip-to-lip surface or a quadrant of the dish to check the declination angle.

The third Arc-Set tool is marked Extreme or No. 3 and is set for the look angle to the lowest elevation satellite. This will be Spacenet 2 in the Western United States and Satcom C1 (F1) east of the Rocky Mountains. Place the Arc-Set tool marked Extreme or No. 3 on the antenna's center plate or a surface parallel to the lip-to-lip surface and actuate the dish down toward the horizon until the tool's bubble is centered. That is the elevation of the lowest satellite. It will be necessary to rotate the tool on the center plate as the dish is rotating down to the lower elevation bird in order to keep the Extreme tool measuring down the steepest angle of slope.

Polar Mount Adjustments

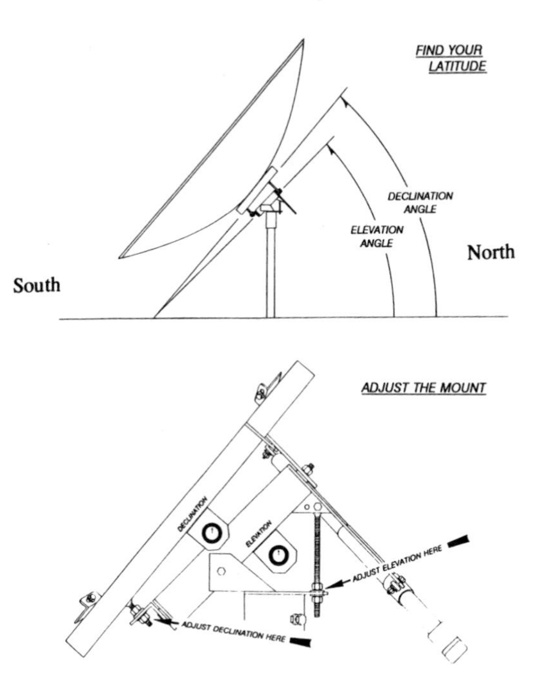

Diagrams courtesy of The Perfect 10 Antenna Company

Note: If an inclinometer is used instead of the Arc-Set tools, it will be necessary to know the exact elevation angle to the lowest satellite from the dish's location.

After setting these three angles, loosen the mount cap bolts that secure the mount to the ground pole enough so that the entire mount can be rotated on the ground pole. While viewing the picture on a TV located at the dish or, better yet, monitoring a signal strength meter connected in line with the LNB, rotate the entire mount on the ground pole until the signal is peaked or at its highest strength. Tighten the mount cap bolts and re-check the bubble and signal level.

Tip: Peaking the signal on the lowest elevation satellite puts the dish on track and locates the true south heading without the need of a compass.

Actuate the dish back across the arc, stopping at a couple of satellites at both ends of the arc and in the middle to check the results. No change in elevation should be required on any satellite if the mount is accurately tracking the arc. Slightly pulling up or pushing down on the lower lip of the dish simulates an elevation change; note if the signal strength goes up or down.

If minor adjustments are required, adjust the axis elevation on the highest satellite (one that is closest to true south) and adjust the true south heading of the mount on the ground pole while monitoring the lowest elevation satellite. Once the mount is accurately tracking the arc, tighten all hardware and scribe a line on the mount cap and ground pole. Use this line as a reference in the event high winds cause the mount cap to turn on the pole and throw the dish off track.

The ability of the mount to accurately track the arc of satellites depends on a number of factors. The ground pole must be plumb. If the concrete base has shifted and the ground pole is tilted east or west, tracking the mount will be difficult.

Today's polar mounts feature bushings or bearings at the axis pivot points and are engineered to provide good stability and smooth rotation. Most are fitted with oil impregnated bronze bushings that require no lubrication. Mounts that rotate on bearings may have a grease fitting and require annual lubrication. Unfortunately, antenna mounts did not always feature bushings or bearings at the pivot points. Some early mounts simply had holes drilled for the pivot points and the dish rotated on bolts. Wear on the pivot points will cause instability in high winds and not allow the dish to maintain the proper elevation throughout the arc.

Even on the best polar mounts, stress created by high winds or heavy snow can loosen hardware and change the mount angles critical to proper tracking. Make sure the mount hardware is tight and check the pivot bushings for excessive wear. This can be accomplished by lifting up on the mount ring or lip of the antenna while the antenna is at its hightest elevation and the weight is evenly distributed on the mount. Be careful not to apply so much pressure on the dish's lip that it distorts the shape.

Rust and corrosion are two of the worst enemies of the polar mount. This is especially true in coastal areas where there is salt air. Some manufacturers offer zinc plating as an option to combat these elements. For iron mounts that have been protected only with powdercoat paint, plan on spending a little time with a wire brush and a can of rust-resistant paint.

—— Horizon-to-Horizon Mounts ——

Motorized horizon-to-horizon (H-to-H) mounts are available as an option by every antenna manufacturer. For long-term, heavy-duty applications, it is a worthwhile investment.

An H-to-H motorized mount integrates the motor and mount in one compact unit, provides antenna movement over 180 degrees and has snow-dumping capabilities.

The Ajak 180 model is the most popular H-to-H mount. It features massive worm and ring gears, built-in limit switches and more than enough torque to move dishes weighing 250 pounds.

The Ajak 180 H-to-H mount costs a little more than a standard mount and actuator, but it is not subject to freeze-ups in cold weather and should provide at least 10 years of little or maintenance-free operation.

While today's polar mounts and actuators will provide years of reliable service, nothing beats an H-to-H motorized mount.

Tip: If a combined C- and Ku-band system is installed, the H-to-H mount will target Ku-band satellites with far greater precision and accuracy than an actuated polar mount.

—— Mounts Q & A ——

Ed. Note: Satellite TV is a complex and constantly changing field. One way dish owners keep up is through the author's "Ask the Tech Editor" column in Satellite TV Week. *Here is a sampling of letters that deal with changes in the last couple of years. They also help give a broader and more detailed understanding of satellite TV.*

Blowin' in the Wind

Q: My 10-foot Unimesh dish was installed last spring in tranquil and sublime weather. The present winter winds are rattling the bejabbers out of it and really giving me such picture problems that it hardly seems worthwhile turning on the set on some days.

Does anyone make and sell a radom-type cover to solve such problems? Should I try constructing a wind-deflecting fence?

I would appreciate receiving any suggestions that might help and hope you don't tell me to move my dish.

— John Mullen, New York

A: *A radom-type cover designed to shed snow and ice or disguise a dish to look like a patio umbrella would only make your problem worse by increasing the wind-load factor. You need to determine what is allowing the dish to move in the wind.*

Every polar mount must be adjusted to track the arc of satellites when a system is installed. Once the adjustments are complete, the installer tightens all the mount hardware and then rechecks the tracking. High winds and heavy snow can cause the hardware to become loose and reduce antenna stability.

Try grabbing the dish by the outer edge and give it a little shake. Look for any evidence of loose hardware at the polar axis, as well as where the actuator arm attaches to the dish and the mount. Tighten all loose bolts and apply lubrication to the polar axis pivot points and the swivel joints on the actuator arm.

Most standard polar mounts offer good stability in mild to moderate wind conditions. That's not to say that the picture wouldn't be lost when the wind is gusting at 50 mph. Like most dish manufacturers, Unimesh offers optional heavy-duty mounts for inclement weather conditions.

Off Track

Q: We have had a satellite system for some time now and enjoy reading *Satellite TV Week*. For about the last eight months, we have had a lot of sparklies and crackling sounds on many channels. We clean up some of the channels with the fine tuning, but not 100 percent.

When baseball season started, we ordered Midwest SportsChannel and the picture was full of sparklies and noise. Fine tuning or auto-peaking only made it worse. The repairmen thought it was anything from a loose wire to a weak LNB or problems with the receiver.

I finally found a repairman with enough time to come out and look at the system and he found the dish was off track. After loosening the bolts and rotating the dish a little to the west, all of the satellites came in perfectly.

— Les Hesley, Iowa

A: *Thanks for the tips. No adjustments to a receiver will compensate for a dish that's unable to accurately target satellites. Feed positioning and tracking should be checked periodically.*

Once the tracking is adjusted and the system is peaked for optimum performance, it's a good idea to scribe a line on the mount cap and the ground pole. If high winds cause the mount to turn on the pole again, all you need to do is align the two marks and retighten the bolts.

Out of Line

Q: I have what seems to be an impossible problem. At this point, I'm grasping at straws.

About four weeks ago, our Chaparral Sierra II had problems. It was overwhelming, causing channels to cut out. It seemed obvious that something was shorting out. Since I had to go to the hospital for two weeks, we unhooked it and took it to a local dealer. The dealer then sent the unit to Chaparral for repair.

When the unit returned, the dealer brought it out and hooked it up. Now I can get G1 and a few other satellites, but not others. The pictures I do get are full of sparklies. The dealer replaced a perfectly good actuator ($200). It didn't help.

I called Chaparral technical assistance and they told me it sounded as if my dish is out of alignment and it takes equipment that I don't have to put it back on track. Is there a company that makes the tools I need? Furthermore, do the tools come with instructions which a layman can understand?

— Gerry Shirley, California

A: *I agree with Chaparral. If you are only getting a few satellites, and even they are bad, the problem is dish alignment.*

The most common cause of sparklies is the inability of the mount to accurately target the dish onto a satellite. Although your dish and mount may have been properly adjusted when first installed, adverse weather conditions, loose hardware and normal wear and tear on bushings require periodic adjustments to keep your system performing at its best. A competent dealer who wants the best for his or her customers always will check antenna alignment to ensure that the maximum signal is obtained from all available satellites in the arc.

Tools required for aligning the mount angles and setting a dish on track vary from one installer to another. I have always used Arc-Set tools from Gourmet Entertaining, 3915 Carnavon Way, Los Angeles, Calif. 90027, 213-666-2728. They can be ordered preset for your location and come with easy-to-follow instructions.

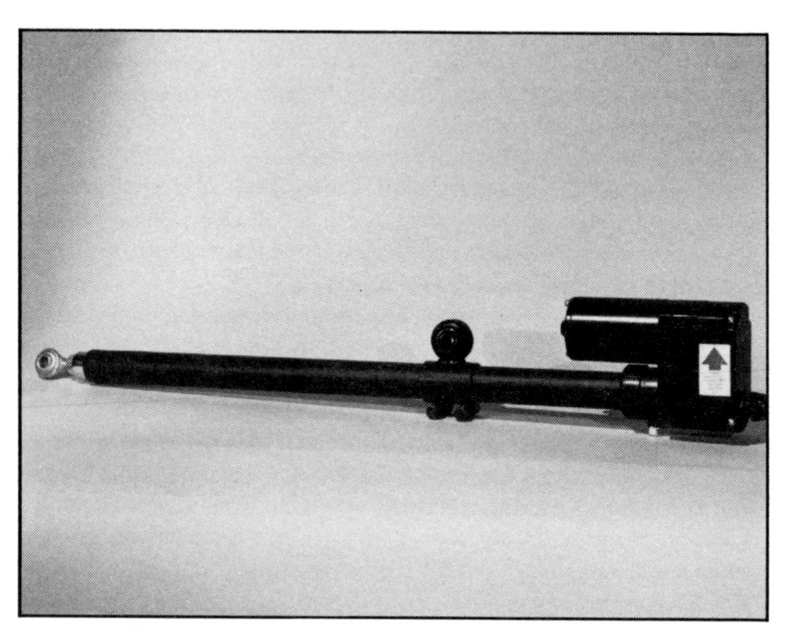

Black Jack Linear Actuator
Photo courtesy of Houston Tracker Systems

CHAPTER 7

Actuators

The purpose of the linear actuator, which attaches between the polar mount and antenna, is to provide stability while targeting a specific satellite and a means of rotating the dish about its axis to access all satellites in the arc.

Linear actuators or jacks consist of a motor and gears with an arm that telescopes in and out of a fixed tube. Acme jack models feature a threaded shaft which moves in a threaded collar and are usually rated for loads up to 500 pounds. Reciprocating ball jacks feature ball bearings instead of a threaded collar. Ball jacks provide smoother movement and are often rated for loads up to 1,500 pounds.

Actuators are available in various stroke lengths from 18 inches to 52 inches. An 18-inch jack is adequate for most 10-foot antennas, allowing the dish to access every satellite over North America. Mounting geometry on larger dishes may require a 24- or 36-inch stroke. Heavy-duty antennas like the old ADM 11-foot models use a tripod mount configuration whereby a 52-inch actuator attaches between the dish and the concrete pad.

Feedback systems located inside the motor housing send information to the controller or antenna positioner located inside the house in order to keep track of the antenna's position. Some early sensing circuits used a potentiometer which sent a varying voltage to the controller. The most common sensor used today is the reed switch which features four or more magnets located on a round plastic wheel which turns with the motor.

The reed switch, located in close proximity to the magnet wheel, is open until a magnet passes by and causes it to close. Each pass of a magnet in front of the reed switch creates a pulse which is sent to the controller. If there are four magnets on the wheel, four pulses will be detected by the controller for every revolution of the motor. Higher resolution sensor circuits are available for more accurate positioning.

Reed switches require only two small wires. One wire is used for ground and the other carries the pulses to the controller. Hall-effect transistors work the same as a reed switch, but require a third wire to supply five volts to the transistor. Regardless of the type of sensor employed, there always will be two heavy wires which supply 24 or 36 volts DC to the motor.

Reed switch failures often are caused by nearby lightning strikes. The switch also can become magnetized and stick closed if all four or more magnets have the same poles facing the reed switch. This can be avoided by turning every other magnet over so that the poles alternate as the magnets pass by the reed sensor.

All antenna controllers, whether a stand-alone component or an integrated part of a satellite receiver or IRD, are designed to shut down when pulses are not received from the sensor located in the motor housing. If the sensor fails or a pulse wire connection is broken, the controller or IRD will usually display ''Jack Error,'' ''Actuator Error'' or something similar. When the east or west keys are pressed, the dish will not move or will move slightly and then stop.

Antenna controllers and IRDs feature electronic limits for the actuator movement which must be programmed into the memory before any satellite positions can be stored. The better actuator motors also feature adjustable limit switches inside the motor housing to prevent the motor from trying to run the arm beyond its mechanical limits in the event the controller's programmable limits fail.

The actuator arm attaches to the mount and one side of the antenna. It's important that the mounting points allow the arm to move the dish about its axis without any lateral force being applied to the arm. This means the arm must be located so that its linear movement is perpendicular to the mount's axis. Use shims or spacers as necessary at the attachment points to prevent any lateral movement of the arm.

The actuator arm is equipped with a swivel joint and clamp that attaches to the rear of the mount and a swivel joint at the end of the arm that attaches to the dish. The swivel joint with the clamp should be positioned on the fixed tube of the actuator so the dish can access the lowest elevation satellite a little before the arm is retracted to its mechanical limit and still be able to target the furthest satellite at the other end of the arc before the extended limit is reached.

By no means should the arm be allowed to extend to a point where it loses leverage when the dish comes around to the opposite side. That could cause the arm to bind against the rear of the mount and bend the arm. Set the mechanical limits in the motor housing so the arm can retract just beyond the lowest satellite and extend just beyond the furthest satellite at the other end of the arc. The arm should be attached so that an angle of at least 30 degrees is maintained between the arm and the back surface of the dish throughout the arc.

Water intrusion is the actuator's worst enemy. All actuators should have drain holes located at the bottom end of the fixed tube where it enters the motor housing and at the lowest point in the housing or back cover. Although water entry between the inner arm and outside tube may be prevented by a wiper boot and internal ''O'' ring when the actuator is new, continued use eventually will wear out the seals. Drain holes are a must. Otherwise, water can accumulate in the tube and motor housing. When the water freezes, the dish will not move.

Preventative maintenance includes keeping the drain holes clear, lubricating the swivel joints and checking for loose hardware. It's a good idea to remove the back cover of the motor housing and inspect the wire terminals for corrosion at least twice a year. If there is rust on the chromed arm, use steel wool or fine sandpaper to clean the surface and apply a little light grease. Protective accordion boots which fit over the actuator arm will help extend its overall life, but like any mechanical device, actuators are subject to wear and eventually will require professional service or replacement.

—— Actuators Q & A ——

Ed. Note: Satellite TV is a complex and constantly changing field. One way dish owners keep up is through the author's ''Ask the Tech Editor'' column in Satellite TV Week. *Here is a sampling of letters that deals with changes in the last couple of years. They also help give a broader and more detailed understanding of satellite TV.*

Universal Remote

Q: I have a M/A-COM T1 positioner; RCA VCR, Model VPT 294; Mitsubishi TV, Model CS 2657R; and a VideoCiper II 2100E descrambler. Is there a ''universal'' remote that will control all of these?

I don't have a remote for the VC II, so I need a remote that does not have to memorize the codes from other remote controls. None of the ''universal'' remotes that I have seen advertised have the ability to handle a satellite system.

I have another problem, too. Since I live in the mountains, we often get high winds and freezing temperatures. When we try to move the dish and then back again when it is frozen, the count goes off. Eventually, I have to clear the memory and start over from scratch. What can we do to keep this from happening?

Thirdly, I do not have the manual for the dish. Is there a way to adjust the arm so we can get more of a sweep than we have? Right now it goes from just west of F1 to just east of F2. This is okay until the counter gets off and then F1 is right on the west limit. It would be nice if we had a little more leeway in adjusting F1 before I have to clear the memory and reset the limits.

— R. Welch, Wyoming

A: *There are a number of universal remote controls on the market that do not need to memorize the codes from your remotes. Instead, these remote controls come pre-loaded with the codes for thousands of products, including satellite receivers. All you do is look up a three-digit code in the code book for each model of equipment you wish to program. If the remote has been preprogrammed for that model, a light will flash and let you know. If it doesn't, you can have the codes entered at an authorized service center. Proton, All-4-One and Radio Shack manufacture preprogrammed remote controls. However, none of them have keys marked specifically for satellites. You'll either have to memorize which numerical keys are associated with each alpha designator or make yourself an overlay for the keypad. Depending on the vintage of your VC II, it may or may not respond to a remote control.*

The problem you are experiencing with the actuator arm freezing up is quite common. There should be drain holes at the bottom of the outer tube where it enters the motor housing and at the lower corner of the motor housing itself to prevent water from accumulating. A rubber boot installed over the inner tube will help to prevent water intrusion, but actuator arms only last for so long. Eventually, the original lubricants get displaced or dry up and freezing temperatures take their toll. Sometimes it's possible to extend the life of an actuator arm by drilling a hole, installing a grease fitting and placing new grease inside the tube. Be sure to use a grease that will displace water and that will not freeze.

The miscount symptoms you describe are probably the result of the clutch slipping inside the motor housing. If the inner and outer tubes of the arm are frozen together or are at the mechanical limit of travel, the clutch is designed to slip. This is a safety feature built into some motors to prevent gear damage. When the clutch slips and the motor continues to turn without engaging the arm to move the dish, pulses from the motor sensor are still returned to the controller. The result is exactly what you have experienced. The actuator count is no longer synchronized with the dish position. Most of the new IRDs include features that allow the user to move the limits without the need to reprogram all satellite posiitons and to resynchronize the counter with the dish position after the clutch has slipped or the motor has been removed for service. Unfortunately, your T1 positioner doesn't offer those features.

The amount of arc coverage depends on the length of the arm and the way it is attached to the dish. Most of the lightweight mesh dishes can cover the entire arc with an 18-inch arm. Some of the early fiberglass dishes require a 24- to 36-inch arm to maintain leverage.

The most practical solution is to install a horizon-to-horizon motorized mount like the Ajak 180. Call Ajak at 303-784-6301 for more information on retrofitting your particular dish.

Changing Numbers

Q: I have an STS MBS-LSR receiver that is giving me problems. I also have a Janeil nine-foot dish.

The actuator numbers for each satellite have changed a bunch. About three months ago, all of the positions changed by three numbers. I had a service man come out and all he did was reprogram all of the satellites, which I could have done myself. Now the numbers are off by 21. I cannot access the satellites and I cannot change the upper and lower limits to reprogram the unit. The west limit was around 19. Now the dish stops at 41. The east limit was 50 and now I've seen numbers as high as 68. Can you please tell me what caused this and how to correct the problem?

Right now I am turning the dish with a crank.

— Robert W. Skidmore, Florida

A: *Counting errors can be caused by a faulty sensor or spurious pulses entering the controller. The most common type of sensor circuit consists of a reed switch and a magnet wheel located inside the motor housing. Every time a magnet passes in front of the sensor a pulse is sent back to the controller. A microprocessor uses the pulses to keep track of each satellite position.*

When a sensor becomes intermittent or fails to send a pulse back to the controller every time a magnet on the wheel passes in front of it, the controller will continue to move the actuator until it receives the programmed amount of pulses for that satellite position. As a result, the controller will display the programmed number, but the dish position will be wrong. If the sensor fails completely, most controllers will stop the actuator and display an error code.

Spurious pulses, such as those caused by power surges or lightning strikes, also can enter the controller. When this happens, the microprocessor can interpret the spurious pulses as sensor pulses and add fictitious counts to the numerical display. From the symptoms you describe, this scenario seems to fit your situation. The use of shielded sensor wires usually will prevent the problem. But since you live in an area that is subject to numerous lightning and thunderstorms, I suggest you take further measures to prevent serious damage to your system. Make sure your system is properly grounded and invest in a good lightning and surge protection device.

You undoubtedly will have to reprogram the STS-LSR receiver. I suggest you start by clearing the memory. The procedure requires removing the top cover and shorting out a couple components to momentarily disconnect the battery voltage from the microprocessor. Unless you're a technician, take it to your dealer and let him or her do the master clear. Then you can take the unit back home and reprogram it. If you need instructions for the master clear procedure or an owner's manual to program the STS-LSR, call STS at 314-421-0102.

SuperJack XL Sensors

Q: I have a SuperJack XL actuator on our system and the Reed sensor keeps going out. Since it was installed in July, it has blown four sensors. My dealer is out of them and no one can give me the name of address for SuperJack. Can you help?

— Elwood R. Johnson, Kentucky

A: *Pro Brand International, Inc., 1900 West Oak Circle, Marietta, Ga. 30062, 404-423-7072, manufactures the SuperJack XL and can supply you with replacement sensors.*

Since you live in an area that is subject to thunderstorms, I suspect that lightning discharges are responsible for the high failure rate. You should consider grounding the dish and installing lightning and surge protection before more serious damage occurs.

Mini-Dishes

Q: There is a 32-inch satellite dish on the market that's "custom made for RVs and the traveling lifestyle." The ad says it performs as well as a 10-foot dish, but I imagine it's for Ku-band reception. Will it work on C-band?

— Frank Braunwart, Kentucky

A: *A 36-inch dish is too small to receive signals from our current C-*

band satellites. The next generation of satellites to be launched will have higher-powered transponders, but I believe that at least a five-foot antenna still will be required. Companies offering mini-dishes are probably counting on the success of proposed DBS ventures.

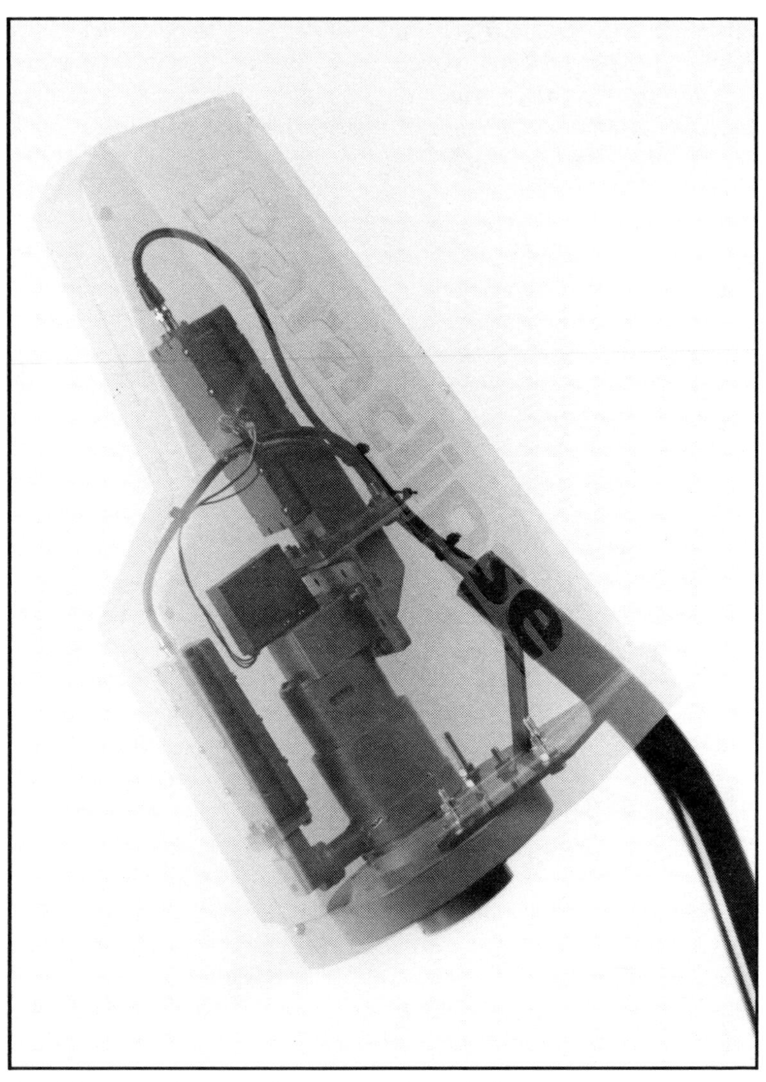

Avoid sharp bends in cable and use appropriate size feedhorn cover
Photo courtesy of Paraclipse

CHAPTER 8
Cables

A special ribbon cable containing all the necessary cables and wires has been specifically manufactured for the satellite industry. These ribbon cables, used to connect the outside components to the inside electronics in today's home satellite systems, consist of two coaxial cables for the LNBs, three stranded wires for the servo motor and five stranded wires for the motor drive. The servo wires and motor drive wires are in separate jackets which are sometimes shielded to prevent noise from entering the line and causing the receiver to pick up false pulses.

Coaxial cable used to connect the LNB to the receiver's input is designated RG-6. Some older (70 MHz) systems used RG-59 which has a smaller diameter center conductor. Although both types of cables are rated at 75 ohm impedance, RG-6 coax can carry signals farther with less loss than RG-59. The "F" connectors for RG-6 are also a little larger than those used for RG-59.

Coax cable is composed of a solid center conductor which carries both the DC voltage and signal, a dielectric foam core, an outer braided shield which also serves as ground, and a protective plastic jacket. The dielectric core establishes the impedance of the cable and serves as an insulator between the center conductor and shield.

Care must be taken when routing coax cable to avoid sharp bends which can change the impedance and cause attenuation on certain channels. If the feedhorn cover doesn't allow for a wide turning radius where the coax cable connects to the LNB, use a 90-degree "F" connector elbow. The same type of attenuation can result if the coax cable is pinched when stapled to a baseboard or floor joists. Staples designed specifically for RG-6 and ribbon cable are available from satellite dealers.

When installing "F" connectors on coax cable, strip the outer sheath back about 3/8 inch from the end, roll back the braided shield and trim the plastic core to expose about 1/4 inch of the center conductor. Be careful not to nick the center conductor. Push the "F" connector onto the cable and use a crimping tool to crimp the ring at the rear of the "F" connector onto the cable. The center conductor should extend beyond the end of threads far enough to make a good electrical connection but not so far that it will cause a short when screwed onto the LNB or receiver input. Never use pliers to crimp "F" connectors. Cable stripping and crimping tools are available to do the job correctly.

Outside cable connections must be protected against water intrusion. A weather cover placed over the feed assembly will prevent direct rain and snow from landing on the feedhorn and LNB, but condensation still can enter the connection. When water enters a cable, it can short out the signal and even damage the LNB or receiver.

Tip: Several methods for sealing outside cable connections are successful. Some dealers use a putty-like tape called Coax-Seal that easily can be molded around the connector. I use a dielectric filler called S.T.U.F. to flood the "F" connector before it is screwed onto the LNB. Both materials are available through your dealer.

The three stranded wires that connect between the servo motor on the feedhorn and receiver are 22 gauge or larger. These are often color-coded red for five volts, white for pulse and black for ground. The motor wires consist of two 14- or 16-gauge stranded wires that supply 24 to 36 volts to the actuator motor. Three smaller wires connect to the sensor in the motor and deliver pulses back to the receiver so that it can keep track of the antenna's position.

Note: Reed sensors require only two wires, one for the pulse and one for ground. Hall effect transistor sensors use a third wire in addition to the pulse and ground wires. The third wire on the Hall effect sensor connects to the five-volt supply terminal.

Plastic ties work great for securing cables and wires at the dish. The LNB cable (or cables in the case of a C/Ku-band feed) and the servo wires can be run alongside one of the feed support legs and around the back of the dish to a place close to the pivot point. Leave a loop in the cables between where the ribbon attaches at the rear of the dish and where it attaches to the pole or enters the PVC weather head. The loop should be large enough to permit antenna movement across the arc without the possibility of the cables getting snagged.

Cable ties also will work to secure the motor drive wires on the lower actuator tube (that's the tube that is stationary, not the arm that extends and retracts). Again, leave enough slack to allow the arm to pivot on the swivel without stressing the wire. Drip loops are necessary where the motor drive wires enter the housing and where all the cables and wires enter the PVC weather head or weather box.

It's good practice to inspect outside cable and wire connections periodically. Regardless of measures taken to seal connections, the elements eventually take their toll. Just remember to unplug the receiver or IRD first. All satellite receivers deliver a constant voltage to the LNB whether the power button is on or off to prevent condensation inside the LNB housing.

Check cable and wire connections for signs of water intrusion and corrosion. If water has entered the cable, use a hair dryer to evaporate the water. Be careful not to apply so much heat that it melts the cable jacket or foam core. Oxidized connectors can sometimes be cleaned with alcohol, but it's better to replace a corroded "F" connector. Motor drive housings have a terminal strip. Clean the connections and apply a little non-conductive silicon grease to prevent corrosion or rust.

Inside connections are not nearly as vulnerable, but wires can become frayed or pulled loose when equipment is moved while cleaning or dusting under components. Tinning the stranded wire ends or installing crimp-on spade terminals will prevent the wires from getting frayed and help ensure good contact. A strain relief can be used to tie the wiring harness to the unit. Always leave enough extra cable and wires so the unit can be moved to clean or service. Make a wiring diagram of both the outside and inside connections and mark all cable and wires with number labels.

—— Cables Q & A ——

Ed. Note: Satellite TV is a complex and constantly changing field. One way dish owners keep up is through the author's "Ask the Tech Editor" column in Satellite TV Week. *Here is a sampling of letters that deals with changes in the last couple of years. They also help give a broader and more detailed understanding of satellite TV.*

Severed Connections

Q: We had our lawn put in and had the satellite TV lead-in cable cut in two. Can this be spliced together or does it have to have the cable replaced?

Will there be any trouble if this is spliced with dampness in the future? You have helped us many times.

— Thomas Bolger, Wisconsin

A: *You can repair the damaged coaxial cable by installing F connectors where the coaxial cable is cut and using a barrel fitting to join the two F connectors together. A terminal strip can be used to join the other severed wires. Waterproofing is a must if you plan on burying the cable.*

I have used a line of 3M products that include a rubber tape (Model 130-C), vinyl tape and a product called Scotchkote, a lacquer coating, in similar situations. Better yet, there are materials available that can form a plastic mold. Check and see what your local electrical supply store recommends for waterproofing underground connections.

Another possible solution is to mount a weatherproof box above the ground, bring the severed cables inside the box and make the connections there. Since the connections would be inside the box, no weatherproofing materials would be required. Of course, this method would require that you have enough slack cable and you don't mind having a little gray box mounted to a post in the backyard.

Oxidation Woes?

Q: I have an STS system with block downconversion and a 75 degree LNA (California Amplifier). Over the past year there have been about four times when the picture degrades (sparklies on the weather satellites and/or transponders). Once I adjusted the feedhorn distance and that cleared it up. Another time I cleaned the connections at the feedhorn and adjusted the feedhorn distance and that did the job. The third time I changed the connector at the downconverter and that did the trick.

On the fourth time cleaning the connectors and drying them out didn't help a bit. Then I remembered that on the three previous occasions I had tapped the box beam holding the feedhorn, LNA and downconverter. So I took my handy crescent wrench, gave the box beam three moderate taps near the feedhorn and, lo and behold, everything was fine.

What's the cause of my problem?

— Frank A. Nowak, Arizona

A: *Once a connector has oxidized, it is best to replace it. If your LNA and downconverter are connected directly together with a coupler, it may be difficult to weatherproof. It is best to use an LNA cable between the LNA and downconverter so that the downconverter can be located behind the dish in a weatherproof box. This procedure will not require you to weatherproof the end of the cable that connects to the downconverter since it will be protected by the box, and also make it easier to disconnect for troubleshooting in the future. If your problem* *still exists, and a change occurs when you tap on the LNA or downconverter, it is possible that moisture has had an effect on one of those components. Oxidation may be causing a problem inside the LNA or downconverter. Do not attempt to disassemble the LNA or downconverter. They should be sent to a repair shop.*

Courtesy of Microwave Filter Company

CHAPTER 9

Terrestrial Interference

Terrestrial Interference (TI) results when a home satellite system receives microwave signals generated from land-based relays operating in the same C-band of frequencies used by satellites. The most common source of TI is telephone company microwave relay towers, although some airport navigational systems also can interrupt satellite reception.

Years ago (before satellites) the telephone companies were allocated the use of frequencies from 3.7 to 4.2 GHz for their common carriers. Since the telephone companies also were the first users of satellites, it only made sense to design the satellites to operate in the same frequency band so that they could use much of the existing technology and hardware for satellite communications.

In order for the satellites to share the same frequency band, it was necessary to restrict the output power of C-band satellites so as not to interfere with the existing land-based microwave communications. Transponders aboard first-generation C-band birds produced five to eight watts of power, requiring a receiving dish of 10 feet in diameter or larger to receive sparkle-free pictures in most of the U.S. The latest-generation C-band satellites have up to 17 watts of power and clean, clear pictures are now possible with dishes as small as five feet in diameter.

However, close spacing between satellites means that reflectors smaller than six or seven feet in diameter may experience interference from an adjacent satellite when targeting a satellite. That is because smaller antennas have a wider look angle at the sky.

Ku-band satellites, in comparison, are not restricted in power. Some current Ku-band birds transmit with almost 50 watts of power, enabling antennas as small as three feet in diameter to receive excellent pictures. The first Direct Broadcast Satellite (DBS) will offer 120 watts of power and beam signals to antennas as small as 18 inches square.

The effects of TI when a home C-band system is caught in the line of fire between two microwave towers talking back and forth can range from light sparklies or a little buzz in the audio to complete picture wipeout. The severity of the interference depends on the level of the interfering signal compared to the level of the satellite signal received by the dish. Some microwave towers broadcast with more than two watts of power, which can create real problems if the offending tower is transmitting signals directly to a home dish from less than 20 miles away.

Telephone relays use both narrowband and wideband transmissions. Narrowband carriers usually support analog traffic and often can be overcome by inserting a narrowband filter. Wideband traffic, on the other hand, is usually digital and cannot effectively be filtered.

Telephone carriers are always offset from the center frequency of C-band transponders by 10 MHz. In other words, if you were receiving TI on transponder 1 (3720 MHz), the interfering carrier would be at 3710 MHz or 3730 MHz, either 10 MHz above or 10 MHz below transponder 1's center frequency. Mild cases of narrowband interference sometimes can be suppressed by off-tuning the transponder frequency away from the interfering carrier at a slight cost in picture degradation. By the way, it is not unusual for one microwave tower to have as many as six carriers.

The best method of combating terrestrial interference is avoidance. That doesn't necessarily mean you have to move to another neighborhood. Often, it is possible to position the dish where some sort of natural shielding, such as a building or trees, can be used to block the offending land-based interference, yet still enable the targeting of satellites over the top of the barrier.

For example, suppose the offending tower is directly south and firing signals right into your dish. You then would want to locate the dish on the north side of the house as close in as possible, without blocking reception from the satellites. Of course, you will first have to know where the tower is located or at least from what direction the TI is coming. That requires getting specific information from the frequency coordinator at the telephone company (highly unlikely) or locating a satellite dealer who is experienced in combating the effects of TI and has the appropriate equipment (spectrum analyzer), which can measure TI levels and identify the offending frequencies and direction from which they are coming. An alternative is to move a portable dish around on the property to locate the site with the least interference.

Proper choice of equipment also can help to reduce the effects of TI. Deep dishes with f/D ratios of .312 or lower have lower sidelobes which is where TI gets into the system. At least one feedhorn manufacturer (California Amplifier) offers feedhorns which feature a scalar designed for dishes plagued with TI.

Many receivers and IRDs feature built-in or optional TI filters which can be engaged on an on-channel basis as necessary. There are accessory filters available for a wide range of applications from several sources. Your best bet is to contact a professional satellite dealer who has experience with TI.

—— Terrestrial Interference Q & A ——

Ed. Note: Satellite TV is a complex and constantly changing field. One way dish owners keep up is through the author's ''Ask the Tech Editor'' column in Satellite TV Week. *Here is a sampling of letters that deals with changes in the last couple of years. They also help give a broader and more detailed understanding of satellite TV.*

Combating TI

Q: We live in a rural area with no cable service and can only receive a very few stations with a regular TV antenna. We have tried several times to install a satellite TV dish with no success. We are located in the path of an AT&T microwave tower. Is there any device we can use to overcome this problem?

— Clyde Jackson, Kentucky

A: *The best defense for terrestrial interference is avoidance. That is, you need to find a location for the dish that will provide natural shielding to effectively block the interference, yet allow reception of the satellite signals. If no natural shielding is available, it may be necessary to construct artificial shielding using a framework and fine mesh screening material.*

The first step is to find a satellite dealer with microwave interference experience and, hopefully, one who has a spectrum analyzer. A spectrum analyzer is used to measure the frequency, level and direction of terrestrial interference carriers so the installer can choose a location for the dish with the least amount of interference.

Your choice of equipment is also important. Deep antennas, with f/D ratios of .32 or less, generally have lower side lobes and better TI rejection. There are feedhorns, such as the Cal-Amp Centerline, that also help to reduce the effects of microwave interference. TI filters are available as built-in features on many IRDs or as accessories that can be added to the LNB cable or IF loop on the receiver.

Microwave Monster

Q: On every scrambled station to which we subscribe, the picture suddenly will black out and then we hear the advertisement. When this happens, the VideoCipher LED will go out. This may last a few seconds or 20 minutes or more. On clear transponders the picture will just blink for a moment. This may be repeated frequently.

I talked to GI and after putting me through a series of tests, they said that there was nothing wrong with the dish or receiver, that the problem may be terrestrial interference. I tried a Phantom IFP adjustable bandpass filter tuned to 70 MHz with no change. I took the receiver to our local satellite dealer and he put it on his bench hooked up to his dish and TV. The system ran rock solid for four days. I hooked the receiver to my system and the problem was still there. I've also replaced the cable, tried three other LNBs and another receiver but no change. Can you help?

— Ken Geist, Washington

A: *From the symptoms you describe, it sounds as if you are a victim of the microwave monster (TI). The reason the filter may not be effective is because the interference is stronger than the satellite signal.*

Your best defense is to find another location for the dish; use natural shielding such as buildings or trees, or construct a microwave shield. Microwave shields are available from New Image Products, Inc. in Portland, Ore., 503-761-9611.

Ma Bell's Towers

Q: I am having problems with terrestrial interference on my TVRO. I'm told it is caused by Ma Bell Communications frequencies disrupting our dish receivers. Can't the FCC do something about these problems?

I have completely updated my system trying to get away from TI. It is tough and expensive to do so.

— Maurice B. Hoyt

A: *Microwave communication links used by the telephone companies operate in the same frequency band as our C-band satelites. In fact, they were established long before any of the communication satellites were launched. Since the telephone companies were allocated the use of the C-band frequencies, power outputs on the satellites had to be restricted to low levels so as not to interefere with their existing com-*

munication links. Until the telephone companies replace their present microwave systems with fiber optics, or satellites start broadcasting in a difference frequency band such as Ku, TVRO installations located in the path of Ma Bell's towers will continue to be plagued with interference.

For more information on how you can combat your interference problems, contact Microwave Filter Company, Inc. at 1-800-448-1666. In N.Y. call 315-437-3953.

HTS Tracker Premiere 70
Photo courtesy of Houston Tracker Systems

CHAPTER 10

IRDs

In the early 1980s, home satellite systems consisted of separate components to tune channels, select polarity, move the dish from one satellite to another and even tune stereo audio programs. By 1985, manufacturers were offering models that combined all these functions in one box and allowed complete operation of the system via remote control. With the introduction of scrambling and the VideoCipher decoder came the first integrated receiver-decoders (IRDs).

IRDs are the brains of today's sophisticated home satellite systems. They include the satellite receiver which is responsible for tuning individual transponders or channels, a power supply for operating the antenna actuator, and a VideoCipher II Plus or renewable security (RS) module for receiving encrypted (scrambled) programs via electronic authorization signals sent by satellite.

The basic function of the receiver in the IRD is to process the downconverted signals from the LNB and extract or demodulate the original video and audio information that was modulated onto the carrier at the uplink. Baseband video and audio signals are made available at the rear of the IRD for direct connection to a video monitor and audio amplifier. These are RCA plug connectors just like those featured on stereo amplifiers for connecting tape decks or other audio sources.

In addition, most IRDs also offer a modulated (VHF channel 3 or 4) output for connecting the IRD to the antenna input on a television tuner. This is the same type modulator used in your VCR. The best pictures and sound result from using the direct video and audio connections between components.

The actuator controller circuitry in an IRD provides power to the motor that moves the dish. It includes a microprocessor that keeps track of programmed satellite locations so the user can recall any satellite by simply keying in the satellite name and number on the remote control.

Initial programming of satellite locations is usually done by the installer. However, system owners should learn how to program their IRDs. Otherwise, every time a new satellite is launched or a satellite moves to a new location, you'll need to contact your dealer for a service call.

Connections for the motor drive include two terminals that provide 24 to 36 volts to the motor and three terminals for the sensor wires. Reed sensors have two wires that connect to the pulse and ground terminals so the microprocessor can keep count of the dish position. Hall effect sensors require an additional third wire to supply five volts to the Hall effect transistor.

Specific instructions for initial programming, changing memory, deleting satellites, adding satellites and selecting satellite formats are outlined in the owner's manual. The actual keystrokes for performing these functions vary from one IRD to another, but before any IRD can be programmed, the polar axis adjustments must be set so the dish is able to accurately track the arc of satellites. In addition, the mechanical limits in the motor should be set so the dish can be actuated just beyond the eastern- and western-most satellites.

Although programming procedures will vary among IRD models, most require the east and west electronic motor drive limits be set before any satellite locations can be stored into memory. These are the programmable limits in the IRD which should be set just short of the mechanical limits in the motor. If the actuator arm has been properly mounted and the mechanical limits set, the on-screen graphics will display "actuator error" or something similar when the dish is moved to the mechanical limits. All you have to do is back the dish up a few counts with the east or west key and store the programmable limit. Once the programmable limits are established, you can move the dish with the east and west keys to find all the satellites.

Most IRDs are set up so you can enter the satellite name and number (G5) before moving the dish in the direction of the satellite. The IRD will automatically select the correct format and an active default channel so the satellite will be easily recognizable when the dish reaches it. Just be sure to use the correct alphanumeric designator listed in the operator's manual.

It's also a good idea to have an up-to-date list of satellites and their longitudinal locations for reference. An excellent source is the centerspread "Channel Choice" chart in *Satellite TV Week,* which is updated weekly. Once a satellite has been located and correctly identified, you will need to adjust the polarity and dish position for best picture and store that information into memory. Most new IRDs feature automatic dish and polarity peaking which makes the task much easier. The same procedure is repeated until all satellites have been located and stored.

One new function featured on some IRDs is an automatic locator. After the limits are set and the first two or three satellite positions are located and stored, the IRD will automatically locate and store all other C- and Ku-band birds.

All IRDs feature audio tuning which allows the user to vary the audio from 5 MHz to about 8.5 Mhz. Normally, the receiver will default to 6.8 MHz on non-scrambled channels because that is where most programmers broadcast the audio associated with the video program. However, many transponders on satellite also have additional audio subcarriers (radio programs) which are not associated with the video on the same transponder. Six or more stereo radio programs are found on some transponders. Some radio programs are broadcast in monaural and others are broadcast in stereo. Reception of stereo audio subcarriers requires an IRD with a stereo processor. Low-end IRDs may only feature monaural subcarrier tuning.

Today's sophisticated IRDs often feature preprogrammed favorite channel menus or menus that permit the user to customize a list of favorite programs for fast, easy recall. Other features may include UHF remote control which allows system operation from anywhere in the home, built-in TI filters and bandwidth adjustments to help combat the effects of terrestrial interference, surround sound outputs, video scan to assist in locating new satellites, programmable VCR timers to permit unattended recording and much more. Of course, every IRD includes a VideoCipher module to receive subscription programming.

There are no user-serviceable parts in an IRD. Just be sure to keep the unit in an area where air can circulate. Overheating can be a problem if the IRD is kept in a closed cabinet. Other than replacing a blown fuse or resetting a circuit breaker, there's little a user can do when an IRD fails.

—— IRDs Q & A ——

Ed. Note: Satellite TV is a complex and constantly changing field. One way dish owners keep up is through the author's "Ask the Tech Editor" column in Satellite TV Week. *Here is a sampling of letters that deals with changes in the last couple of years. They also help give a broader and more detailed understanding of satellite TV.*

Multi-receiver System

Q: We recently moved from Orange County, Calif., to Rogue River, Ore., and are renting a house with a satellite system while we build a house.

Our family consists of our 27-year-old handicapped daughter, my 92-year-old mother-in-law, my wife and me. Because of this, we have four distinct viewing habits.

Is there a system available (other than four separate dishes) that will allow us to view four stations at a time?

I have noticed the cable operators in two local cities broadcast multiple channels from one large antenna system. Is there a system available for household use that will accomplish this?

Also, is there a VCR that will position the dish and record a program at a scheduled time?

— Douglas R. Leger, Oregon

A: *You can connect multiple satellite receivers to one dish and tune programs independently at each receiver. However, the dish can only target one satellite at a time. That means you all would have to choose programs on the same satellite.*

Cable companies use multi-beam antennas which pick up signals from more than one satellite at a time, but they are very expensive.

I suggest you make a list of the programs each member of the family likes to watch and then check the listings in Satellite TV Week. *You will find that movies on the premium channels are repeated often enough in any given week or month that there are plenty of opportunities to see the same movies. In addition, movies on channels such as HBO, Showtime and Disney, as well as the networks, are broadcast at least two different times a day to accommodate various time zones.*

If you have the available space, four independent systems will give you the most flexibility. The cost for the extra electronics to allow multiple receivers to operate from one dish will be about the same price as another dish. You easily can combine signals from satellite receivers, VCRs, laserdisc players and off-air or cable channels so that all sources will appear on every TV in the house.

Most top-of-the-line receivers include programmable timers that move the dish and select a channel for up to 14 different events, but you still have to program the VCR.

Clearing Memory

Q: I have a Uniden UST 9900 and the dealer I bought it from is no longer in business.

When I am reprogramming my receiver, which I have to do about four times a year, is there a way to erase the memory? I accidentally programmed a few numbers and letters into the receiver such as T9 and G9 and would like to reprogram it without the mistakes. Unfortunately, I've lost my owner's manual, so any information would be helpful.

— Marty Williams, Pennsylvania

A: *The UST 9900 does not really provide a simple method for master clearing the memory. However, you can erase individual satellite locations by moving the dish to the satellite you wish to erase and holding the memory key for about four seconds or until the display flashes.*

In order to master clear all memory, it is necessary to go inside the receiver. The procedure should be performed by a qualified technician. First, unplug the power cord and remove the top cover. Locate the yellow battery on the top control board. There will be a two-pin jumper covered by a black sleeve next to the battery. Unplugging the jumper for a period of two minutes will cause the microprocessor to lose all memory. Master clearing the memory would require that the limits, satellite locations and channel-specific parameters such as fine tuning and audio frequencies would have to be reprogrammed, just as if the receiver had never been programmed.

You can write to Uniden Service Center, 9900 West Point Dr., Indianapolis, Ind. 46250 for a new owner's manual.

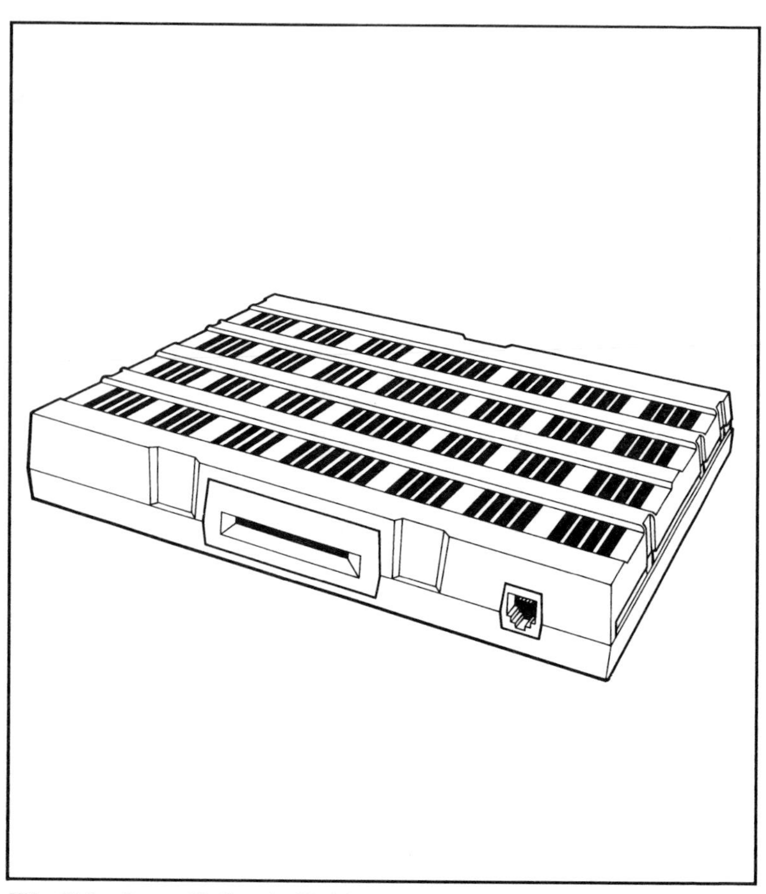

VideoCipher Renewable Security Module
Courtesy of General Instrument

CHAPTER 11
Decoders

The General Instrument (GI) VideoCipher decoder module was first introduced in a stand-alone unit that could be interfaced with popular satellite receivers. These stand-alone decoders featured their own power supply and connected to satellite receivers using either the 70 MHz or baseband interface. A few receivers proved to be incompatible without some modifications. Today's IRDs incorporate a space to accommodate the VideoCipher module and power is supplied to the module by the IRD.

The first-generation VideoCipher II decoders were compromised almost immediately after their introduction. That is, hackers found ways to modify the decoders to receive programming without authorization.

Piracy spread like wildfire. The decoder which GI had thought to be secure had been defeated. GI made one revision after another on its production line and continually sent out electronic countermeasures in an effort to thwart the "chippers," but piracy continued to plague the industry. Programmers were losing hundreds of thousands of dollars in subscription fees. Something had to be done.

Several years later GI introduced the VideoCipher II Plus module which features a higher level of security and offers more tier bits, or room for more programmers. As of this writing, security of the higher tier bits has yet to be compromised by the hacker underground.

But GI didn't stop there. GI knows that chippers will eventually crack the VC II Plus codes as well. Thus the latest version of the VideoCipher II Plus decoder is the RS which stands for renewable security. It features a slot that is designed to accommodate a plug-in card that can be used to upgrade the security of the VC II Plus module.

Supposedly, if programmers feel the VideoCipher II Plus is compromised to the point that enough revenues are lost, they will be able to issue a TvPass Card to subscribers with RS modules. The consumer just plugs the card in to update the security. However, not everyone has a VideoCipher II Plus RS module and nobody has ever seen a working TvPass Card in a RS module. To be fair, the RS module was introduced in early 1992 before a break in the VC II Plus and before there was a need for a TvPass Card. Also, GI says it can produce one within months after a break occurs, and that it will tailor the card for conditions at that time.

In 1992, GI and the programmers undertook a major task to upgrade owners of older VideoCipher II modules to the newest VideoCipher RS module. Only those with legitimate (untampered) VC II modules and active subscriptions as of March 31, 1992, qualified.

Overall, the upgrade program was a huge success, pleasing all segments of the home dish industry except the pirates. Programming customers who qualified for the program received a free VC RS module. If they did not feel comfortable installing the new module themselves, GI paid a service fee for the local dealer to come out and install the module.

There were some owners of older VC II modules that fell through the cracks and for various reasons did not qualify for the free upgrade. Their module may never have been authorized by the March 31, 1992, cutoff date. Others may have let subscriptions lapse or moved without notifying their programmer of the new address. Unfortunately, there also were some cases of incompatibility where the RS module was blamed for causing power supply failures in the IRD.

The upgrade allowed programmers to discontinue using the datastream that addressed the older consumer VC II decoder modules and move all their services to the VC II Plus. This move rendered older VC IIs which had been modified to receive programming without authorization useless. Presently, only customers with VC II Plus or VC RS modules can be authorized to receive programming. Programmers no longer can authorize the older consumer VC II modules because that datastream has been discontinued. However, a certain piracy element still exists.

Although the consumer VC II datastream which the pirates once depended on has been discontinued, a commercial VC II datastream still exists on all major services. It accommodates commercial VC II decoders used by cable companies and the hotel/motel market. Hackers are somehow getting the commercial codes used to authorize the commercial VC II decoders and modifying the consumer VC IIs to respond to the commercial codes. Eventually, all commercial VC II decoders will have to be upgraded to commercial VC II Plus decoders. There may even be a commercial VC RS decoder.

The Federal Communications Commission (FCC) is conducting an investigation into the home satellite decoder market. Up to this point, GI has had no competition. But that is in the process of changing. At least one other manufacturer, Titan Satellite Systems, claims to have a compatible decoder. If it can persuade programmers to authorize its Linkabit SCS modules, there will be competition in the decoder hardware marketplace and a lowering of decoder prices. At press time, that appeared to be happening.

Reception of subscription programs with a VC II Plus or VC RS module requires a good signal. Otherwise, the decoder may have trouble authorizing. If you tune a non-scrambled channel when the dish is a little off target, you may get a watchable picture with some noise or sparklies. The same low-level signal conditions on a subscription (scrambled) channel can cause the decoder to cut in and out or fail to authorize for reception. Whenever problems are encountered on a subscription channel, select a channel in the clear, peak the dish and polarity for best picture and then turn back to the subscription channel.

An example of the diagnostic data screen for the VC II Plus and VC RS modules is shown at the end of this chapter. The screen can be viewed by pressing "Setup" and then "0."

The Hit Counter is the first four digits on line 7. This number can be cleared to zero by entering "0" while viewing the diagnostic data screen. You should see an increase in this number on the first working day of every month when the programmers send authorization messages or when you call and request a re-hit.

The Good Frames Count is the third group of four digits in row 7. It also will clear to zero when "0" is pressed, but will immediately start counting upwards. Ideally, this number should be increasing at a rate of 60 to 70 frames every 10 seconds.

The Audio Holds Field in line 8 indicates the ability of the descrambler to successfully descramble the signal received. If this number is anything other than "00," then the descrambler is having difficulty descrambling the signal received. This difficulty could either be caused by a signal problem (S:S with 03, 04, 05, 0A) such as a noisy picture; or missing authorization (S:NS, S:MP, S:OC).

—— Decoders Q & A ——

Ed. Note: Satellite TV is a complex and constantly changing field. One way dish owners keep up is through the author's "Ask the Tech

Editor" column in Satellite TV Week. *Here is a sampling of letters that deal with changes in the last couple of years. They also help give a broader and more detailed understanding of satellite TV.*

Intermittent VC II

Q: My satellite system keeps cutting in and out on scrambled stations. I called my programmer because I thought I might have missed a payment. They informed me that I was in good standing and had me bring up the diagnostic data screen. When the program came in, it blinked a couple of times and went back off again.

The person I talked to said there was a battery in the set that had to be replaced by my dealer. He said other channels were not affected because they are free.

The last time I had a problem of intermittent VideoCipher operation, my dealer came out, charged me $56 and didn't fix the problem. Can I put this battery in myself? If so, where can I buy one and where does it install? My unit is a GI 2400R IRD.

— Paul Costick, New York

A: *The most common cause of intermittent VideoCipher operation is a weak signal resulting from antenna mispointing, a bad LNB or bad cable connection. If you're getting generally poor pictures across the arc, I'd suspect one or more of the above. Otherwise, you've got a problem with the IRD or VideoCipher module. You may want to call General Instrument at 704-327-2026. The information that appears on the diagnostic data screen will give them a clue as to what's wrong.*

The battery in the module is a backup system for the unit ID number in the event of a power failure. If the battery were to fail, you would lose the ID number the next time there was a power failure or you unplugged the IRD. When this happens, the ID number usually will turn to all zeroes and the module is considered dead. Replacing the battery at that point would neither restore the unit ID number nor its operation. When a module "zeroes out," all you can do is purchase a new module.

Balky VideoCipher

Q: I have a Houston Tracker SRD-8000 IRD. For about a year, I have had an interesting problem. When first turning on my unit to an encrypted channel, the picture would flash on and off rapidly or would turn black even though the VideoCipher light was on. If I pushed the reset button on the back of the unit it would clear up the problem. It would happen once every few days or weeks.

Recently, the problem has gotten worse. Almost every time I turn on my unit, I must push the reset button to get an encrypted channel to come in. If I change channels or satellites, I have to reset it again.

In addition, none of my controls related to the VC II operation (setup, next program, help, messages) work even if the VideoCipher light is on and the picture is coming in. I have no way to view the diagnostic or setup screens at all.

— Paul Vitale, Tennessee

A: *SRD 8000 IRDs have had some problems with a couple of plastic plug connections due to heat generated inside the unit. The connections become intermittent and cause the decoder to cut in and out. By pressing the reset button, you're probably just shock exciting the module into working by the sudden change in voltage.*

The cure for the problem requires hardwiring around the plastic plugs, but should be done only by a competent repairman. Have your dealer call the HTS technical hotline at 303-790-7878 for details.

The modification should restore normal VideoCipher operation, including the setup functions. If not, have your dealer try another module.

Pay Per View

Q: I see movie listings each week for Cable Video Store, Action and Viewer's Choice 1 and 2, but I don't see how to order these movies. Is the process similar to TVN? If so, what phone number do I call?

— Pete Kaeding, Kansas

A: *The movie listings in question are pay-per-view services available to dish owners with a VideoCipher II Plus and stand-alone Videopal order/recorder or a VC II Plus module with a built-in order/recorder. The new VideoCipher RS module has a built-in Videopal. The Videopal is a modem that connects between your decoder or IRD and telephone line. It is available through local satellite dealers. Once connected, you can call the Satellite Video Center at 1-800-54-VIDEO and set up an account. Pay-per-view programs or events then can be ordered by simply pressing a key on your remote control.*

All TVN programming also is available to VC II Plus owners without a Videopal by dialing an 800 number. However, you first must establish an account with TVN by calling 1-800-232-4TVN.

Viewer's Choice 1 listings at 8 p.m. and 10 p.m. (E) are available to VC II Plus owners without a Videopal through HBO at 1-800-285-8856.

Receiving VC II Authorizations

Q: I need a better understanding of how scrambling works. I have a Sierra III with an integrated VC II and I subscribe to some scrambled channels on G1 and F3. In understand that I need to get an authorization signal every now and then for those to continue to work.

I have now discovered SCOLA on S2/16, and recommend that every satellite viewer check it out. It is an extremely valuable source of information, especially if you, like me, moved here from Europe.

After watching SCOLA, is it necessary that I move the dish down to G1 or F3 to receive my authorization signal, or can I park the dish on the nearest scrambled channel even if I do not subscribe to that particular channel to get my authorization?

Or is it possible to predict when an authorization signal is sent? If so, I would only have to remember to move the dish to one of my scrambled channels on those days.

— Kenneth Axelsson, Georgia

A: *Authorization messages are sent on the first working day of each month. The messages can be received on any scrambled channel using the VideoCipher scrambling system.*

All integrated receiver/descramblers (IRDs), such as your Sierra II, automatically search for a scrambled VC II channel on whatever satellite the dish is parked when in the standby mode (power off). Those with stand-alone VC IIs need to leave the receiver on and tuned to a scrambled channel to receive the messages. In either case, if you miss the authorization message, you can call your programmer for a direct hit.

VCII Plus/VC RS Diagnostic Data Screen

V1.07

1) 02324544FFEF
(Unit Number-Do not use this number when ordering programming. Use Unit Number on Setup 1-1 screen.)

2)	3:0000	2:0000	1:0000	0:0000
3)	000000	000000		
4)	00	XXX-YYY-ZZZ		0E-1
5)	0000-0000	0000-0000.00		0000
6)	C0-3E-30	0	5046	0-0-000000
	(DBS Service ID (CO)		Monthly Epoch (3E)	Channel ID) (30)
7)	0006-0931-0454		4-0-01	000000
	(Hit Counter (0006)	DBS Messages (0931)		Good Frame Count) (0454)
8)	0.00E-00/2.00E-06		00	S:S
			(Audio Holds)	VC II Plus Module Status

Common VC II Plus Module Status

First Letter S = Full Scrambled F = Fixed Key
P = Processed Unscrambled

S:S = Unit is authorized. You should be receiving picture & sound.
(Same as SA on VC II)

S:NS = You are not subscribed to this channel.
(Same as SB on VC II)

S:MP = Unit has never been authorized or is tuned to CATV only channel.
(Would appear as SM on VC II)

S:OC = Unit has not received monthly update or unit is defective.
(Would appear as SM on VC II)

S:CB = Program is blacked out in your area.

S:SL = Program is locked out due to rating ceiling.

HTS MasterMind
Photo courtesy of Houston Tracker Systems

CHAPTER 12
Accessories

Today's modern home entertainment systems may include a satellite receiver, cable TV or outside antenna to receive local programs, several TVs and VCRs, laserdisc player, security camera, camcorder and a full complement of audio components.

Such complex systems need a convenient method of switching and distributing the various signal sources to different TVs, monitors and recording devices. In this chapter we will explore some of the innovative accessories which allow you to combine and distribute signals, view one source while recording another, control devices from remote locations and more.

Video sources such as satellite receivers, VCRs and video disc players have two different output signals available for connection to a monitor, TV or audio amplifier. The audio and video outputs from such components are suitable for direct connection to the audio and video inputs of a VCR, monitor or audio/video amplifier. RF (radio frequency) signals are also available for direct connection to TV and VCR tuners. The RF output usually can be sent to VHF channel 3 or channel 4 on a TV set or monitor. Using both the A/V (audio/video) inputs and outputs as well as the VHF inputs and outputs of components provides the flexibility needed for recording one source while viewing another.

The simplest of home entertainment systems will include a satellite receiver, outside antenna or cable, TV and a VCR. In order to view either a satellite signal or an off-air channel while recording the other, you need a three-way RF switch and a two-way splitter.

The three-way switch is fed by the RF outputs of the satellite receiver, the VCR and the VHF antenna. A two-way splitter is used to feed the RF signals from the antenna or cable to the VCR and the three-way switch. RCA cables are used to connect the A/V outputs of the satellite receiver to the A/V inputs of the VCR. Note that some VCRs may have switching inputs when using the A/V jacks. If so, it may be necessary to disconnect the A/V cords to record off the regular TV antenna. Be sure to use a high-quality three-way RF switch that features enough isolation between inputs to prevent interference. See figure 1.

An alternative is to combine the off-air signals and VHF channel 3 or 4 output from the satellite receiver so that both signals are present at the VCR tuner and TV antenna terminals simultaneously. Generally, only VHF channel 3 or 4 will be broadcast locally. This means you can set the modulated output of the satellite receiver to the unused local channel.

However, you can't simply use a two-way splitter backwards to combine the off-air VHF channels with the modulated VHF output of the satellite receiver because VHF modulators in satellite receivers and VCRs produce unwanted sidebands which will interfere with the adjacent VHF channel. Channel combiners are available for under $20 which are designed to accept either VHF channel 3 or 4 from a satellite receiver and the VHF channels from an off-air antenna. These special combiners eliminate co-channel interference and work fairly well as long as the inputs are balanced.

Distribution Systems

Multiplex Technology, Inc. is one company that offers a complete line of multi-room distribution products. Information is available by calling 1-800-999-5225. Its ChannelPlus modulators accept audio/video inputs from sources such as satellite receivers, VCRs, cameras and disc players, as well as VHF off-air or cable channels, and combine the signals so the audio/video source simply appears on an unused off-air or cable channel.

The combined signals then can be distributed to all TVs and VCRs in the house. Off-air or cable channels are tuned on their regular channel positions. The audio/video source such as the satellite receiver is viewed by tuning to a user-selected, unused UHF or cable channel on the TV or VCR tuner. ChannelPlus modulators are available in models to accommodate up to three separate A/V sources.

Vu-Tech Communications, Inc. has introduced SelecTeVision Plus, a new multi-room entertainment system that allows viewing of VCR, cable TV and/or satellite on up to three different TVs in the home. With just one VCR and one IRD, you also can tape one program on the VCR while watching a different program and use your existing remote controls for your VCR, satellite receiver, cable converter and audio components from any room where a SelecTeVision remote set top unit is located.

The system contains a small set-top unit for each television set which transmits control signals back to the control unit. The "Main Set Top Unit" for use at the television set where the converter (or IRD), VCR and cable feed (or antenna) are located is used to select which source is to be viewed and what source is recorded at the main TV location.

For further information, contact Vu-Tech Communications, Inc., 100 Circle 75 Parkway, N.W., Suite 800, Atlanta, Ga. 30339, 404-850-1590.

—— Wireless Transmitters ——

Distribution systems like the ChannelPlus and SelecTeVision Plus send their signals via coaxial cable to each TV and VCR. There are a number of wireless transmitters on the market as well. These systems use radio waves to send signals from a transmitter located near the source to a receiver at the remote TV location. But picture quality varies considerably among models. The best I have seen is the VideoPath, which uses FM transmissions. VideoPath broadcasting systems are available from satellite dealers as well as many video stores.

—— Remote Controls ——

Now that we've looked at ways to distribute signals from the satellite receiver, VCR and off-air antenna to TVs and VCRs throughout the home, we need a method of controlling those devices from the remote TV locations.

Virtually all IRDs feature on-screen graphics. Many IRDs also feature UHF remote control which permits system operation from anywhere in the home. But how can you change channels on your satellite receiver from another room with an infrared (IR) remote? Furthermore, how can you operate other IR remote control components such as your VCR from another location?

VideoLink offers a line of products which extends the range of infrared remote controls so IR devices can be operated from remote locations using the original equipment's remote control. The product line includes infrared receivers, power supplies, senders, interfaces, connecting blocks and cable extensions.

Some of the VideoLink products are designed for custom installation directly into walls. The entire house can be wired so that you can control any infrared device from anywhere in the home. Other models allow remote transmitters to relay IR signals to components inside closed cabinets and to send IR signals over existing coaxial cable. VideoLink products are available through satellite dealers and custom audio/video installers.

In addition to infrared extenders that rely on wires or coaxial cable to send IR control commands back to components from remote locations, there are several wireless systems available that convert the IR signals to UHF and back to IR.

The Infrared (IR) Wireless Remote Control System 2000 from ChannelPlus enables the homeowner to control a VCR, laserdisc player, satellite receiver, or other video source located in another room. The same type of wireless infrared remote extender also is marketed under the X-10 Powermid label. Used with a ChannelPlus modulator, you can watch and control any video source from any room in the house.

With the MasterMind from HTS, it is possible to control all of the infrared-controlled entertainment products from anywhere in the house with one remote control. The MasterMind works by memorizing the commands from all infrared remote controls and transmitting those commands via UHF (Ultra High Frequency) to the MasterMind's "transponder" which in turn converts the UHF signals back to infrared and delivers the signals to the sensor in the unit you wish to control.

Surge Protection

A satellite system represents a substantial investment and should include protection against power- and lightning-induced surges. Even static charges from lightning strikes a half mile away can have a devastating effect on the sensitive electronics in an LNB or satellite receiver.

Inexpensive AC surge protectors often prove inadequate. It doesn't make sense to use a $29.95 surge device to protect a $2,900 investment, especially if you live in the southeast where lightning and thunderstorms are a common occurrence in the summer months.

Lightning-induced surges can enter a system via the LNB cable, polarizer wires, motor sensor and control wires, cable TV or rooftop antenna and phone lines connected to a PPV modem, as well as the house AC lines. All lines where power problems can damage a satellite system must be protected. In addition, the entire system must be properly grounded as outlined in the chapter on installation.

Panamax is one company that has designed lightning and surge protection devices specifically for home satellite systems. Its latest product, the Towermax/Sat, protects all cables and wires and includes a lifetime warranty that covers the device and any properly connected equipment. Similar devices are offered by Electronic Specialists, Evergood Fabrication, Jayco Electronics, Polyphaser Corporation and Universal Technology and are available through satellite dealers.

—— Pay Per View ——

There are two ways to purchase pay-per-view (PPV) programming: via 1-800 call ahead or with a General Instrument Videopal.

The 1-800 system requires no accessories other than a nearby telephone. You call the 1-800 number, the operator or computer records your phone number and decoder unit ID, a signal is sent out over satellite to authorize your decoder, and a picture appears on your TV screen.

The Videopal comes in two forms. It can be either a separate, stand-alone box about the size of two stacked remote controls, or it can be a modem built into the VideoCipher II Plus or VideoCipher RS module. All VC RS modules, by the way, have a built-in Videopal, but that's not true of all VC II Plus modules. All are available through satellite dealers.

The Videopal is essentially an order recorder that is connected to your phone line and your decoder-equipped satellite receiver. It is authorized initially by calling the Satellite Video Center. From that point on, all you have to do to order a pay-per-view movie or special event is to press buttons on your stand-alone decoder box or your IRD's remote control. At a later date, the Videopal is connected by phone to the Satellite Video Center and reports your PPV purchases. This produces a bill which is mailed to you by the Satellite Video Center.

Stand-alone Videopals have to be plugged into a separate power supply, the IRD and a phone jack. Built-in Videopals only need to be connected to a phone jack.

—— Accessories Q & A ——

Ed. Note: Satellite TV is a complex and constantly changing field. One way dish owners keep up is through the author's "Ask the Tech Editor" column in Satellite TV Week. *Here is a sampling of letters that deal with changes in the last couple of years. They also help give a broader and more detailed understanding of satellite TV.*

Signal Distribution

Q: We have three TVs in our home. While taping from satellite, I am able to view the program on all three sets. But in the playback mode, viewing is only possible on the primary TV set. This doesn't seem normal. Might I be doing something wrong?

The system consists of a 10-foot dish and a Star Trak 8 IRD with a VideoCipher II Plus and Videopal. The VCR is a Hitachi brand.

— Albert Smith, Florida

A: *You probably are dividing the signal from the output of the IRD and distributing it to the primary TV set via the VCR and directly to the second and third TVs. In other words, the VCR only is connected to the main TV. If you connect the RF output of the IRD to the antenna input on the VCR and then distribute the output of the VCR to all three TV sets, you can either view a satellite program or a videotape on all three TVs. However, this configuration will not allow you to view satellite on one TV and a video tape on another TV simultaneously.*

When you have multiple sources such as a satellite receiver and VCR, as well as a local antenna or cable TV, the ideal distribution system will allow you to view any channel or source on any TV in the home. Just such a system is possible with a Channel Plus modulator. Channel Plus modulators allow you to add audio/video sources such as satellite receivers, VCRs, videodisc players and security cameras to an existing off-air or cable TV system and have the additional sources appear on unused UHF or cable channels on every TV and VCR in the home simultaneously.

For more information on Channel Plus modulators, contact your local dealer or call 1-800-999-5225.

KO'd by Lightning

Q: I have been reading *Satellite TV Week* for a few years now and am very impressed with all the fine information you have put out. It's like taking a course in TV electronics.

Now I have a question. I have a Luxor 9534 antenna controller and a Luxor 9550 satellite receiver. The receiver works beautifully. The antenna controller has been knocked out of commission either by a line surge or lightning activity. Is there anyone in California who can repair the unit?

— Charles Cotel, California

A: *Pacific Satellite and Electronics was the original Luxor repair center on the West Coast. You can contact it at 1401 Lincoln Ave., Napa, Calif. 94558, 707-226-7488.*

Adjacent Channel Interference

Q: When viewing the picture from my satellite receiver, it is virtually never clear. Frequently there are diagonal lines across the screen, similar to heavy corduroy. The lines are present the majority of the time. They are always straight and evenly spaced but can range diagonally from one direction through vertical in the other direction. They are never horizontal.

I also should tell you, when I am watching normal TV on channel 3 (the satellite is received on channel 4), the picture is not sharp and clear. Channels 6, 10 and 13 are always extremely clear. Do you have any suggestions?

— William J. Sweeney, California

A: *The symptoms you describe are classic examples of adjacent channel interference. Channel 3 from your outside antenna and the channel 4 modulated output from your VC II Plus are interfering with each other. Homestyle modulators, like those used in VCRs, satellite receivers and VideoCipher decoders, usually have low output powers and do not have built-in bandpass filters to limit their output to the targeted channel. These low-cost modulators produce unwanted sidebands that interfere with adjacent channels.*

One way to overcome the interference is to use a high-quality A/B switch with at least 40 to 50 dB isolation between ports. However, this means you'll need to get up out of your chair every time you want to switch from satellite to local antenna or vice versa.

A more convenient method is to use a high-quality channel combiner. The combiner accepts two inputs, an off-air line from a standard TV antenna and either channel 3 or 4 modulated satellite signals. High-quality combiners include a bandpass filter that cuts out the lower sideband and restricts the satellite TV input to a selected narrow range of frequencies. The output is a combined signal. In order to use a channel combiner, it's important that the inputs are balanced, so it may be necessary to amplify the modulated output of the satellite system before connecting it to the combiner.

For the best recording results from satellite, connect the audio/video outputs of the decoder to the audio/video inputs on the VCR and use the "camera" or A/V selector switch to send the satellite signals directly to the recording head. If your TV also is a monitor, connect the A/V outputs of the VCR to the A/V inputs on the monitor. This method may require you to leave the VCR power on, but it will provide the best viewing and recording results and eliminate interference from the modulator.

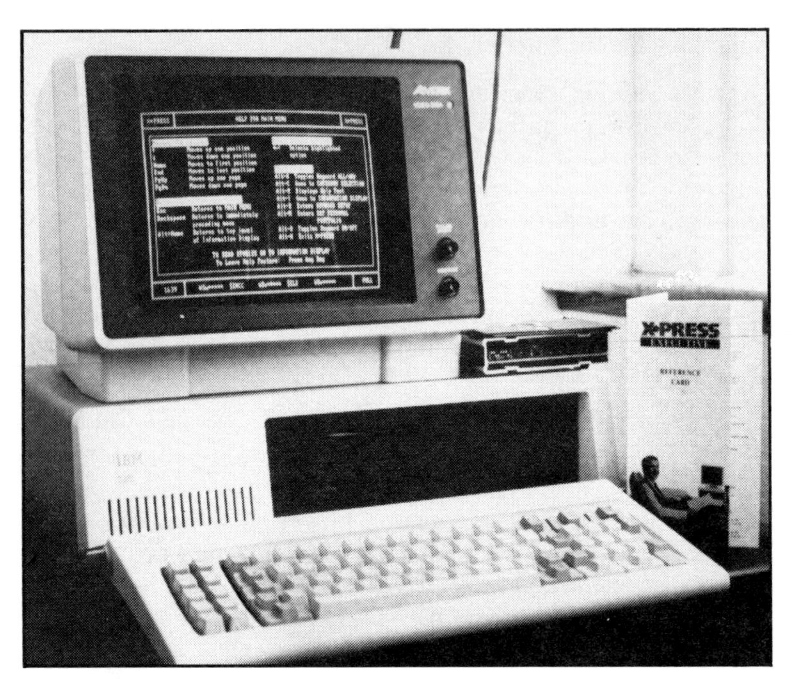

X*Press delivers a constant stream of the latest worldwide news and information
Photo courtesy of X*Press

CHAPTER 13
Text and Data

Welcome to the information age. Besides offering the most choices in entertainment, a home satellite system also provides instant access to an ever-changing world of news, sports, weather and business information, all transmitted in the form of digital data.

The power of satellite technology has changed the way we look at ourselves and the world in which we live. Imagine having a constantly updated stream of late-breaking news stories, scores of games in progress and changes in the financial markets appear on your television or home computer screen. Several affordable devices now on the market can link digital data to your television or home computer, virtually turning your home into a newsroom loaded with wire services from around the world.

The World System Teletext (WST) is one format of video text (interactive data service) that inserts information into an unused portion of the outgoing video signal called the Vertical Blanking Interval (VBI). Coded in the form of digital data, the teletext information doesn't interfere with the regular video but can be recovered and displayed with the aid of a teletext receiver or "decoder." The decoder reads the coded data from the VBI of the picture and decodes it for display on a TV screen in the form of full-color printed data and graphics.

Until recently, teletext information only could be received and displayed using one of Zenith's Digital System III TVs that features a built-in teletext decoder or by purchasing a commercial teletext decoder.

Astro Products Co., 757-8 Twin Oaks Valley Rd., San Marcos, Calif. 92069, 619-471-9930 now offers the Teletext 90 decoder at a suggested retail price of $279. As an add-on accessory, it is connected and operated much the same as a cable TV converter/descrambler, making available either the regular broadcast or the teletext display on a given channel.

Electra is a text and data service. It uses the VBI on WTBS to deliver information in the form of an electronic newspaper to cable systems and dish owners. Top stories of the hour are displayed in headline form with a page reference for further information. For the sports enthusiast, fast-breaking updates include game previews, team ratings and scores of matches in progress. Weather, TV listings, business and general family information also is included on the menu.

Tempo Text is another service found on WTBS that features an electronic stock market quote constantly updated throughout the day with closing prices displayed overnight until trading resumes the next business day. Stock market quotes are delayed from the actual trading by 15 minutes.

The best part about having a teletext decoder is that there's no monthly charge for the services it can receive. It's like having an electronic newspaper delivered every 20 minutes.

Whether you're a serious investor, pin-striped business executive, or just an info-maniac, General Instrument's InfoCipher 1500 Data Receiver offers even greater amounts of information through a service called X*Press. This system also broadcasts on WTBS, as well as CNN, but it uses a data channel on the VideoCipher II Plus instead of the VBI.

You'll need a satellite system that includes an integrated receiver/descrambler (IRD), the InfoCipher 1500R Data Receiver, X*Press software and associated cable kit which connects to your personal computer. The InfoCipher 1500R Data Receiver, X*Press software and cable kit is available from X*Press (1-800-7PC-NEWS) or your local General Instrument satellite dealer for a suggested retail price of $349.95. Monthly charge is $10 for X*Press and $26.95 for X*Press Executive which has a 15-minute delay and contains options and additional news.

InfoCipher and X*Press software give the user a constantly updated stream of news from Associated Press, Standard & Poor's, McGraw-Hill, TASS (USSR), Xinhua (China), Kyodo (Japan) and Agence France Presse before they make the evening news or morning paper. Stock feeds can be accessed from NYSE, AMEX, QTC, NASDAQ, Toronto, Montreal, Vancouver and Alberta with Standard & Poor's Personal Portfolio.

If that's not enough, commodity market information is transmitted three times a day. The highs and lows of the world's precious metals also can be monitored, plus updates on interest, money, exchange and treasury rates.

Another way to get satellite-delivered information is through the Datasat box from DMSAT, P.O. Box 68, Capitola, Calif. 95010, 408-464-2301. The unit can download information directly into your personal computer.

DMSAT sends data via a satellite audio subcarrier instead of using the VBI or VideoCipher data channel. This allows you to record the data on a conventional cassette tape or VCR and play it back at your convenience. There's a receive-only model for $349 and a receive/transmit model for $379. The receive/transmit model enables you to send data via a telephone modem to DMSAT and share comments with others via an electronic bulletin board; the user also can record a message on an audio tape or VCR and mail it to DMSAT.

Data Transmission Network (DTN) is yet another electronic information service that offers an agricultural package with market quotes, grain and livestock information, weather, and updates for all major commodity exchanges. A Wall Street package contains information on financials, stocks, government bonds, currencies, metal, commodities and interest rates. DTN transmits its time-sensitive market information on C- and Ku-band.

If you have a satellite dish with a diameter of eight feet or larger that receives Galaxy 1, transponder 3, the C-band unit will work for you. DTN provides its own terminal, so you don't have to tie up your PC, for a monthly fee of $24.95 (billed quarterly) and a one-time start-up fee of $225.

If you don't want to use your C-band dish, the small (30-inch dish) Ku-band system is the answer. The Ku-band system comes complete for a monthly fee of $29.95 and a one-time start-up fee of $295. A new color version of the Ku-band system features weather radar images and leases for $44.95 a month. Contact DTN, Embassy Plaza Building, 9110 West Dodge Rd., Suite 200, Omaha, Neb. 68114, 402-390-2328.

If you've ever seen the "CC" or little square box with a tail at the end of program descriptions in a TV guide or at the beginning of a TV program, you may have wondered what these symbols mean. The "CC" means that the program has been closed captioned by the National Captioning Institute (NCI).

NCI developed the closed captioning technology that permits hearing-impaired viewers to read the dialogue and narration of television programs with the use of a special Telecaption decoder.

Although the service was originally designed to help the hearing-impaired viewer enjoy television programs, the Telecaption decoders also help children learning to read and individuals learning English as a second language. Being able to read the dialogue and narration at the bottom of the screen is also useful in loud environments where it is difficult to hear the audio.

Closed caption decoders are distributed through Sears Catalog and Service Merchandise. In addition, NCI has a network of more than 1,200 retailers who distribute Telecaption decoders. For more information you can call NCI at 1-800-533-WORD.

Within a year, all TVs sold in the U.S. will include built-in captioning capability thanks to a bill signed by former President George Bush.

Note: All prices quoted in this chapter were current as of press time. Contact the appropriate companies for current prices, equipment and service details.

—— Text and Data Q & A ——

Ed. Note: Satellite TV is a complex and constantly changing field. One way dish owners keep up is through the author's "Ask the Tech Editor" column in Satellite TV Week. *Here is a sampling of letters that deals with changes in the last couple of years. They also help give a broader and more detailed understanding of satellite TV.*

Data Receiver

Q: I would like to take the stock market quotes that I see coming across the bottom of my TV and enter this information into a computer.

I do not have a receiver to do this and do not know what type I need or where to purchase it. Time-delayed quotes, because they are cheaper, are acceptable for now. Later, perhaps, I may be willing to pay more for "real time" quotes.

Any information you can provide on the type of receiver and where to purchase it would be appreciated.

— Danno Anderson, Washington

A: *The General Instrument InfoCipher Data Receiver is available through satellite dealers nationwide. It is a device that connects your VideoCipher decoder and your computer.*

Via the InfoCipher, you will receive a constantly updated stream of data from AP, UPI, Canadian Press, McGraw-Hill, TASS, Standard & Poor's, Zinhua and Agence France Presse news services. Stock feeds from NYSE, AMEX, QTC, NASDAQ, Toronto, Montreal, Vancouver and Alberta also are available. The service is broadcast on

*WTBS and uses X*Press software.*

*X*Press also has programs that will allow you to analyze the data in all sorts of ways, creating charts and graphs, even setting up alarms to alert you to crucial price changes. If you're a heavy PC user, software such as Microsoft Windows will allow X*Press to run in the background so you don't have to tie up your PC while collecting data. Delay time is about 20 minutes.*

*In addition to news and stocks, the InfoCipher and X*Press software will give you weather conditions around the world, tell you what's hot at the movies, and even what your horoscope says.*

For more information on the InfoCipher Data Receiver, see your local satellite dealer or call General Instrument at 704-327-4700.

Closed Caption

Q: We bought a Zenith Digital TV last year and discovered that the Teletext on pay channels such as Showtime and HBO would display closed captions. However, this does not work on over-the-air broadcasts such as ABC.

Will the Zenith TV display closed captions for the networks by going to another page of Teletext? Why are their formats different?

— Richard Hartman, California

A: *The Teletext decoder that is in your Digital System 3 television is designed to decode text information that appears on lines 17 and 18 of the unused and unseen portion of the television signal known as the "vertical blanking interval" (VBI). Caption data is on line 21, field 1 of the VBI. The reason you are able to pick up captions on some of the movie channels is because they are probably dual feeding the caption information on line 17 and 18 as well as line 21. However, the "de facto" standard for closed captioning has been established for line 21. If you wish more information about closed captioning or wish to find out where you can purchase a Telecaption 3000 decoder, call the National Captioning Institute at 1-800-533-WORD.*

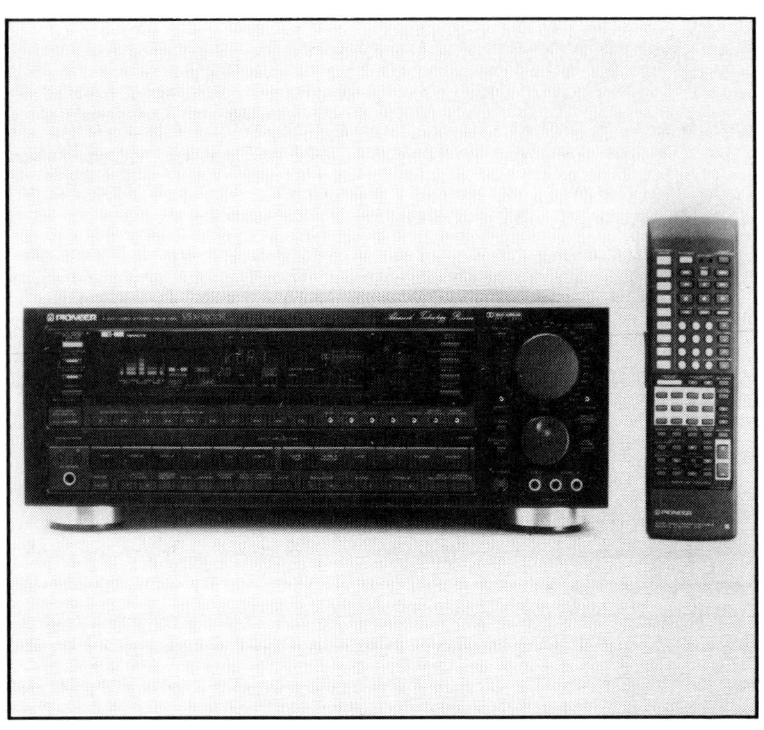

Pioneer A/V Receiver w/Dolby Pro Logic Surround Sound
Photo courtesy of Pioneer Electronics

CHAPTER 14
Audio

Besides offering an incredible variety of viewing entertainment, satellite TV also hosts over 100 radio programs. You can choose from classical, country, easy listening, jazz and rock music, as well as news and information services, religious programs, talk shows and much more.

The audio portion of a satellite signal is transmitted as a subcarrier which is impressed or piggybacked onto the video carrier. For most channels in the clear, the audio which accompanies the video is tuned at 6.8 MHz.

Some networks broadcast two or three audio subcarriers on channels in the clear to provide left and right audio channels for stereo reception, as well as a subcarrier for monaural. However, many more audio signals other than those simply associated with the primary video information can be transmitted in the audio subcarrier band on the same channel or transponder.

By narrowing the audio bandwidth, as many as 10 additional audio programs can be broadcast on the same transponder. That could be 10 different mono programs or five stereo pairs.

The primary audio subcarrier associated with the video program is generally broadcast in a wideband mode. Subcarriers used for audio-only radio programs are most often narrowband, especially if there are several radio programs on the same transponder.

All satellite receivers and IRDs allow the user to vary the audio tuning from about 5 MHz to 8.5 MHz and most receivers have a narrowband filter for accessing the auxiliary radio services. Stereo satellite receivers have two audio subcarrier processors for tuning the two, discrete (left and right) audio frequencies on stereo broadcasts. Audio programs are listed in the "Channel Choice Services" chart in the centerspread of *Satellite TV Week*.

In order to tune an auxiliary radio service like Classical Selections, you first select the satellite and channel (G5/21). If your receiver's audio is preprogrammed or tuned to 6.8 MHz, you'll hear the sound associated with the video on Mind Extension University. Next, switch to narrowband and tune the audio to either 6.3 MHz or 6.48 MHz. If your receiver has a stereo processor, set the left audio to 6.3 MHz and the right audio to 6.48 MHz. It's that easy.

The VC II Plus or VC RS decoder will automatically select the digital audio on subscription channels. Although most radio subcarriers are on non-scrambled channels, there are a few on scrambled channels as well. Some IRDs will allow you to tune audio subcarriers on encrypted channels by simply selecting the audio mode. Others may require additional keystrokes to bypass the digital audio associated with the video.

It should be noted that, with the exception of two IRD models, the VHF output (channel 3 or 4) is monaural. To hear a program in either digital stereo (VC II Plus/VC RS channels) or subcarrier stereo (non-scrambled channels) you must have the left and right audio outputs of the IRD connected to the left and right audio inputs on a stereo monitor or amplifier.

The quality of audio on subscription services like HBO is excellent. But those tiny speakers in some TV sets can hardly do it justice.

If you have a stereo amplifier and speakers, the amplifier probably has a set of auxiliary inputs. All you have to do is connect the left and right audio outputs of your IRD to the left and right auxiliary inputs on the amplifier and arrange the speakers so they're positioned on either side of the TV. Speakers which are not shielded for use in video systems will have to be placed at least a foot away from the TV. Otherwise, they may distort the color.

If you haven't had the pleasure of listening to your satellite system through a good pair of stereo speakers, you're missing the real impact of today's movies.

Those who don't own a hi-fi system or want to upgrade their antiquated equipment will want to check out what's new. Stereo audio receivers of past years, which featured built-in AM/FM tuners and inputs for phonos and tape decks, have evolved into audio/video receivers designed to function as the control center in today's complex home entertainment systems. These extraordinary devices provide inputs for every audio/video source, outputs for a video monitor and a simple method of switching between sources, as well as the capability to record one source while monitoring another. Many models also include Dolby Pro Logic surround sound processing to usher the aural effects of the movie theater into the home.

—— SCPC Audio ——

Many of the National Public Radio stations, Mutual Radio Network, sports networks news and syndicated radio programs are distributed via satellite to selected stations across North America using a form of broadcast known as Single Channel Per Carrier (SCPC).

SCPC signals are very narrowband FM signals that cannot be tuned with a standard satellite receiver. Your home satellite receiver or IRD can tune many of the audio subcarriers found on satellite, but the narrowband SCPC signals are lost when processed by the wideband discriminator in an IRD.

In order to tune SCPC radio programs, you'll need a special SCPC radio. Currently, there are two companies which offer SCPC radios for consumers, Heil Sound (618-295-3000) and Universal Electronics (614-866-4605).

SCPC radios are designed to work with any high-block (950-1450 MHz) IRD. The SCPC radio simply connects to the LNB cable. Another short coaxial cable connects between the SCPC radio and the LNB input on the IRD. Neither of the two SCPC radios available feature a built-in speaker, so you'll need to run the line output to an auxiliary amplifier or amplified speaker.

There are over 200 SCPC radio programs on satellite, the majority of which can be found on Galaxy 2 and Galaxy 6. These very narrowband carriers stand alone on transponders which have no video carriers. As many as 50 SCPC radio programs can be broadcast on a single transponder.

—— Audio Q & A ——

Ed. Note: Satellite TV is a complex and constantly changing field. One way dish owners keep up is through the author's "Ask the Tech Editor" column in Satellite TV Week. *Here is a sampling of letters that deal with changes in the last couple of years. They also help give a broader and more detailed understanding of satellite TV.*

Digital Radio

Q: I have two questions. First, is the music video channel "Video Jukebox" available to the home satellite market? I've read that many cable systems carry this channel, so logically it should be on a satellite. Next, I've read of a new radio service based out of Atlanta called "Digital Music Express." It broadcasts 30 channels of music to cable systems via satellite. Could you please tell me more about these two new services? Can I receive them?

— Richard Comport, Oregon

A: *Video Jukebox Network, Inc. doesn't distribute its service via satellite. Instead, each cable company has its own "Video Jukebox" installed at the cable headend.*

Digital Music Express, Digital Cable Radio and Digital Planet are three companies that have been involved in delivering digital quality or simulated digital sound to cable companies via satellite. Cost for the headend equipment runs around $8,500. A special tuner that costs between $100 and $200 is installed at the subscriber's home.

Although cable-delivered audio services aren't a new concept, adding CD-quality sound might turn this old idea into a new business. One satellite receiver manufacturer was going to incorporate Digital Planet reception equipment into its IRDs, but Digital Planet folded before that could happen. The other two companies have not announced any plans to make their services available to the home dish market.

Recording Surround Sound

Q: Thank you for your help through the years. I hope you can give me an answer to a very short question.

If I record a show off satellite that is in surround sound, will I be able to record the surround information on my stereo VCR, or will the recording just be in stereo?

— Ernie Cypert, Arizona

A: *In Dolby Stereo, in addition to left and right channels, there is a center channel to sharpen the perspective of the on-screen sounds and a surround channel to immerse the listener in ambience and special effects. The center channel and surround channel information is encoded in the film's two audio tracks.*

When the two encoded channels of a Dolby Stereo film are transmitted on television, or transferred to videocassette or videodisc, the stereo signal still carries the center and surround channel information. As long as you use the direct audio/video connections between your stereo satellite receiver or IRD and the stereo VCR, the encoded center and surround information will be recorded on the tape. If you play the tape back through a conventional stereo system, the sound will be two-channel stereo. A Dolby Surround processor will decode the surround information for the rear channel, but does not decode the center channel information. Dolby Surround Pro Logic decoders feature a discrete center channel output and include circuitry which enhances the directional information to improve separation between all the channels.

Audio Translator

Q: I have a Toshiba TRX-120 IRD and the audio range does not permit me to tune lower than 5 MHz. I have asked my dealer about this, but he's been unable to find an accessory that would increase the audio tuning range. Who can I call to find out what I need? Is there such an accessory available? I need something that will tune down to 1 MHz.

— Randy Faldon, Arkansas

A: *United Satellite Systems, Box 351, St. Hilaire, Minn. 56754, 218-681-5616 offers the T-1 Translator which will allow you to tune from 0 to 9 MHz. However, you also will need a separate stereo processor because there's no way to mount the T-1 Translator inside your IRD and connect it to the inboard stereo processor. The T-1 ($95) connects to the baseband or audio subcarrier output on your IRD. A stand-alone stereo processor then can be connected to the output on the T-1 Translator. You may be able to find a used stereo processor at a local satellite dealership. If not, you can purchase one of United Satellite Systems' SSP-1 stereo processors for $75.*

Country Music Fan

Q: Can you please tell me how to pick up country music on my satellite dish? I always notice country music radio listed in the centerfold of *Satellite TV Week*, but don't understand how to tune these radio programs.

— Royce Malone, California

A: *Our home satellite receivers are designed to detect audio and video information spanning from near zero to about 10 MHz. The video portion occupies from near zero to 4.2 MHz. This leaves the rest of the band for audio subcarriers.*

The most common audio subcarrier frequency used to accommodate the video program on unscrambled channels is 6.8 MHz. Stereo broadcasts use two audio subcarriers, one for the left audio channel and one for the right audio channel. Some satellite receivers can only tune one audio subcarrier at a time, or one monaural channel. Others include a stereo processor which enables the user to tune two separate subcarriers for stereo sound.

Check the radio listings under the Country category and you'll find America's Country Favorites on G5/21, followed by D-N and the two audio subcarriers (5.04 and 7.74). The D-N means that the program is broadcast in discrete stereo (two subcarriers) and that it is narrowband.

Your owner's manual will tell you how to select between wideband and narrowband audio and how to vary the audio frequency to tune various audio subcarriers.

Tuning Audio Subcarriers

Q: We have a Uniden UST 7000 satellite receiver and a General Instrument 2100 E stand-alone descrambler with a VC II Plus board. The audio and video outputs run through an RCA VCR and then connect to a Sansui G 4500 stereo amplifier and RCA TV.

Before we purchased the VC II Plus, we enjoyed listening to radio broadcasts on satellite, namely, WCCO in Minneapolis-St. Paul. Since connecting the VC II Plus we only get the sound carried on scrambled channels regardless of what audio frequency is tuned on the satellite receiver. I read something about how to rewire the system in your excellent publication at one time but cannot locate it. Is there a convenient way to eliminate the descrambler in order to tune audio subcarriers on scrambled channels?

— John Mizerka, Wisconsin

A: *The reason you are unable to tune audio subcarriers on scrambled channels is because the decoder is locked onto the subscription program. You need a method of bypassing or looping the signals around the decoder so you can tune subcarriers on scrambled channels. There are two ways this can be accomplished.*

Currently, the audio outputs of your satellite receiver are connected to the audio inputs on the decoder and the audio outputs of the decoder are connected to your stereo amplifer.

Install a "Y" connector on the left audio output of the satellite receiver. Plug one leg of the "Y" connector into the left audio input on the decoder and the other leg of the "Y" connector into one side of an A/B switch. The left audio output on the decoder connects to the other side of the A/B switch and the output of the A/B switch connects the left audio input on the stereo amplifier.

Do the same for the right audio channel. This will allow you to use the two A/B switches to select either the direct audio outputs from the receiver or the audio outputs from the decoder.

An even simpler method is to leave everything connected just as it is and use the small switch labeled "Input" on the rear of the decoder to turn the decoder on and off. The "Input" switch selects either the 70 MHz or baseband input of the decoder. If your satellite receiver is connected via the baseband, setting the switch to the 70 MHz side will turn the decoder off and allow the audio output from the satellite receiver to loop through the decoder. If the satellite receiver is interfaced with the 70 MHz on the decoder, switching to the baseband position will accomplish the same thing. Just make sure the satellite receiver and decoder are not connected by both the baseband and 70 MHz.

The only drawback of using the "Input" switch on the rear of the decoder is that it may not be convenient to reach. In addition, it's a very small switch and not designed for constant use.

Surround Sound Via Satellite

Q: We have a Uniden 9000 receiver with a stand-alone VideoCipher II Plus. Can we utilize an Onkyo TX-SV70 audio/video amplifier to receive surround sound transmissions via satellite? If so, how would we connect the Onkyo to our present setup?

I am building an entertainment room in my house and want to incorporate surround sound.

— Edward R. Harmon, Florida

A: *Your new TX-SV70PRO receiver includes a Dolby Pro Logic processor and will work great with your satellite system. Just connect the left and right audio outputs on the rear of the VC II Plus to a pair of audio inputs on the rear of the Onkyo audio/video receiver.*

If your TV is also a monitor, connect the video output on the rear of the VC II Plus to the video input on the Onkyo and the monitor output on the Onkyo to a video input on your TV/monitor. That way you can use your new Onkyo A/V receiver as the main switching center for all of your audio/video sources.

MTS vs. Discrete

Q: When I watch a show that is in stereo, I do not receive a stereo signal. The stereo indicator does not light up at all on my TV, which is fully capable of receiving stereo signals. When I rent a movie at a video store and watch it on my VCR, the stereo sound is magnificent. Can you please tell me why I don't receive a stereo signal when watching premium programming on TV? Is it possible that my TV is not compatible with the discrete signal on satellite?

I have a 46-inch Hitachi Projection TV, Sears MTS stereo VCR and a Cheyenne IRD.

— Lester L. Keim, Ohio

A: *Multi-Television Sound (MTS) is a transmission system that was developed by Zenith and adopted by the industry in 1984. Today, many of the nation's TV stations broadcast programs with MTS stereo sound. Consumers who are within receiving range of one of these stations can tune to the off-air signal with a conventional rooftop antenna and a TV or VCR equipped with an MTS tuner. The MTS stereo indicator on the TV or VCR only lights when you are receiving an off-air MTS stereo signal.*

Satellite broadcasts do not use the MTS transmission system and will not cause the MTS stereo indicator on your MTS stereo TV or VCR to light.

In order to receive and record satellite programs in stereo, you must use the direct audio/video connections between your satellite receiver, stereo VCR and stereo TV monitor. The VHF output on your satellite receiver and VCR are mono.

Your Hitachi project television is also a monitor. That is, it can accept direct audio/video signals from a satellite receiver, VCR, laser-disc player or video camera as well as off-air or cable channels via the antenna inputs. Typically, projection television/monitors will feature one or more antenna inputs and one or more A/V inputs with the ability to select between various sources with the remote control. There also may be direct A/V outputs on the back of the projection TV suitable for recording.

Baseball on SCPC

Q: We have noticed that Fred Hoffman often gives reference to something called SCPC, noting that the audio for all baseball games is available to those who own an SCPC receiver. What is SCPC and how can we find out more about it?

— Jay Adams, Alabama

A: *Single Channel Per Carrier (SCPC) is a method of broadcast used by many of the nation's Public Radio Stations, Mutual Radio Networks, sports networks and news stations. The majority of SCPC signals are found on Westar 4 and Galaxy 2. Unlike audio subcarriers, which ride piggyback along with video programs, SCPC signals are transmitted on transponders which are not used for video.*

Heil Sound, located at No. 2 Heil Industrial Blvd., Marissa, Ill. 62257, 618-295-3000, offers the SC ONE SCPC radio that works with any hi-block (900-1450 MHz) receiver or IRD. The SC ONE connects to the LNB via a two-way splitter. Tuning is done with a 10-turn pot on the front panel. Price is $450.

Raspy Audio

Q: Since Anik E2 has come on the air, the audio on most channels from my Drake 324 is raspy and slurry. I can't tune out the distortion in either the wideband or narrowband modes. Would a new Drake IRD help? What can I do?

— E.J. Stuart Jr., West Virginia

A: *Most domestic satellites broadcast the audio which accompanies the video program with a bandwidth of about 280 KHz. I believe the problem you are experiencing with the raspy sound on Anik E2 may be the result of an even wider bandwidth. Few new IRDs offer an audio bandwidth greater than 300 KHz, but Drake happens to be one of them. Both Drake and Chaparral Monterey IRDs allow the user to select from three audio bandwidths up to 600 KHz. I suggest you visit a local Drake or Chaparral Monterey dealer and listen for yourself.*

Home Theater
Courtesy of Shure HTS

CHAPTER 15
Home Theater

Films like *Star Wars*, *The Abyss* and *Raiders of the Lost Ark* feature stupendous sound tracks that create such a compelling atmosphere in a movie theater that you respond emotionally and become part of the story. If you ever had the pleasure of seeing one of these action-packed films in a good theater and then later rented the tape and played it at home, you know it's just not the same.

There are several factors at work in a movie theater to bring about the cinematic illusion. First, there's the large screen that fills your peripheral vision with the director's wizardry. But it's not just the size of images that evokes the senses. It's the Dolby stereo sound track that makes the Starfighter seem to come from behind, roar over your head and then suddenly appear on the screen. Atmosphere also plays an important part. The theater is dark with absolutely no ambient light to distract the audience.

Ushering the visceral impact of a theater experience into the home requires a large-screen TV, Dolby surround or Dolby Pro Logic surround sound system and controlled lighting. Laserdiscs offer outstanding pictures and sound, but a satellite system is the only source of live entertainment that can produce high resolution pictures with digital stereo and surround sound.

A true home theater should include a large viewing screen. Custom installers and designers recommend at least a six-foot diagonal screen. Of course, that means a two-piece projection system which not everyone can afford. The alternative is either a rear-projection TV or large TV monitor.

Unlike conventional, two-channel stereo, Dolby stereo films offer four channels of audio. They are the left and right front channels, a center channel to articulate dialogue, and a surround channel to create ambience. The center channel and surround channel information is encoded in the left and right front channels. In a theater, a special Dolby stereo processor is used to decode these channels.

"Dolby Surround" is a term applied when a Dolby stereo film is transferred to videocassette, laserdisc or broadcast. Processors designed to decode Dolby surround sources can be put into two basic categories:

• A Dolby Surround decoder is a passive device that contains a matrix circuit to recover the surround channel from a two-channel, Dolby-encoded stereo source. Dolby Surround decoders restore the vital front-to-back dimension, but do not decode the center channel.

• A Dolby Pro Logic decoder is an active device that uses a variable matrix circuit and offers additional steering to improve separation between all channels. Pro Logic decoders also provide a true center channel output to improve the localization of on-screen sounds and allow a wider image.

Surround Sound processors vary considerably in performance and price. Less expensive processors often offer poor steering geometry, produce irritating clicks and pops in the rear channels or allow dialogue to leak into the surround speakers.

Dolby Surround or Dolby Pro Logic processors can be purchased in many different configurations. Some are sold as a stand-alone device designed to be used with discrete amplifiers. There are also models which feature built-in amplifiers for the center and rear channels, whereby you can use your existing stereo amplifier and speakers for the left and right front channels and purchase additional speakers for the center and surround channels.

Virtually every major audio manufacturer, including Pioneer and Sony, offers audio/video receivers with built-in Dolby Pro Logic surround processors. Typically, these A/V receivers will feature 70-watt or better main channels and 15- to 25-watt center and surround channels.

The dynamic range of sound recorded in today's movies can be extremely demanding on speakers. If your present speakers don't have the capability to handle deep bass, you may want to look at some of the powered subwoofers designed specifically for video.

The ultimate in home theater sound is THX. Designed to re-create in cinemas the exact soundscape that had been generated for films such as *Raiders of the Lost Ark*, THX implies stringent performance standards for signal processing, balanced amplifiers and matched speakers.

The technology has been licensed to several hi-fi companies, including Technics and Fosgate. Both offer complete THX-certified home systems.

—— Home Theater Q & A ——

Ed. Note: Satellite TV is a complex and constantly changing field. One way dish owners keep up is through the author's "Ask the Tech Editor" column in Satellite TV Week. *Here is a sampling of letters that deals with changes in the last couple of years. They also help give a broader and more detailed understanding of satellite TV.*

Stymied by Numbers

Q: Talk about confusion?

What exactly does the term "horizontal lines of resolution" mean? I know only that it is a measure of picture detail.

I thought that American TV used the NTSC 525-line standard. Yet in one of your recent replies to another writer, you stated that S-VHS' 400-line resolution was superior to broadcast TV's 330. In the same letter, you also stated that "normal" VHS uses a 240-line format. I've also read that HDTV is supposed to have a 1,000-line resolution. What's going on? Where are the lines disappearing to? Are we getting "something for nothing" using HDTV?

Could you take the time to explain how these "line-count translations" occur? I think a lot of us "techies" would like to know.

— John C. Cole Jr., New York

A: *The scan rate of American televisions is 15,738 lines per second, which translates into approximately 525 lines drawn every 1/30 of a second. However, not all of the lines are visible. Some are blanked out as the beam traversing the screen returns to the starting pont. Others, at the top of the screen, are used for timing pulses and special information such as closed-captioning. As a result, a typical broadcast picture or "frame," which is drawn in two interleaved passes from top to bottom, consists of no more than 340 lines. This is known as vertical resolution.*

Horizontal resolution is a count of discrete elements that can be discerned along each scan line across the screen, not the number of lines you see when you get close to the screen. That's vertical resolution and is determined by the NTC-SC standard.

Horizontal resolution is determined by the bandwidth of the video signal and the way it is processed. The picture tube and its components also play a part in defining the quality of the image. The NTSC broadcast has a bandwidth of 6 megahertz, of which 4.18 HMz is used for picture information. In our standard TV sets, each MHz of broadcast bandwidth permits about 80 lines of horizontal resolution. So a good TV can reproduce a substantial portion of an NTSC signal, or about 334 lines.

Monitors with specifications of 500 or more lines of horizontal resolution feature video circuits with increased bandwidth in order to accommodate sources such as video disc players, S-VHS camcorders and Extended Definition (ED) Beta which reaches in excess of 500. A standard VCR can record only about 240 lines of horizontal resolution. That's why, when you record a program on a VHS VCR and play it back, the picture never looks as good as the original broadcast. On the other hand, a Super VHS VCR can capture every bit of an NTSC broadcast and render a picture that's identical to the original broadcast when played back. This is possible because the S-VHS VCR records the video signal's brightness component over a wider, higher frequency range on improved Super VHS videotapes.

Improved Definition TV (DTV), another term you probably will hear if you are shopping for a new high-resolution TV, applies digital techniques to our NTSC scanning standards by either storing the first field in memory until the second field comes up and then joins the two together or by adding synthetic scanning lines to those that are broadcast. I have seen some of these new sets demonstrated at trade shows and was able to detect motion artifacts that appeared to streak the picture.

True HDTV, with its 1,000 or more scanning lines and increased width, is still a few years down the line. It will take the FCC at least a year to test the proposed systems before a standard for broadcast can be established and manufacturing can begin.

APPENDIX I

Satellite Talk

Every industry has its own special terms, and satellite TV is no different. It's not necessary to be able to talk satellite to use a modern satellite system. But an understanding of the most commonly used terms helps when talking to a satellite dealer, reading an owner's manual or chatting with a friend about all that high technology in your back yard.

Here is a glossary of the terms home satellite dish owners are likely to encounter:

Actuator
Motorized device used to position a satellite dish for reception of programs. Actuators are built into a horizon-to-horizon mount; they look like motorized shock absorbers when attached to a polar mount.

Antenna
Parabolic dish designed to collect electromagnetic signals from a satellite.

Audio Subcarrier
Carrier wave that transmits audio information.

Automatic Frequency Control
Circuit that locks onto a frequency to eliminate drifting off channel.

Auto Tracking
IRD (integrated receiver-decoder) feature that automatically locates and stores all satellite positions into memory.

Auto Tuning
IRD feature that automatically adjusts the dish position and antenna polarity for best picture.

A/V Switching

Feature that allows users to connect one or more sources such as a VCR, camera and/or laser videodisc player and select which source will be monitored.

Azimuth-Elevation (Az-El)

Antenna mount that allows dish movement in both a horizontal plane and vertically in elevation to locate satellites.

Backhaul

Term applied to the satellite signal sent back to a TV station when a sports "home team" is playing a game on the road.

Baseband

The basic direct 6 MHz output signal from a television camera, satellite receiver or VCR.

Bird

Nickname for a satellite.

Block Downconversion

A process of lowering the entire band of frequencies in one step to an intermediate range. Allows receivers in a multiple receiver system to independently select channels on a satellite.

Buttonhook

A rod shaped like a question mark that supports the feedhorn and LNA. Feedhorns and LNAs also can be supported by three- or four-leg mounts affixed to the edges of the satellite dish.

C-band

The 3.7 to 4.2 GHz band of frequencies, which is the dominant mode of satellite broadcast to the home dish owner.

Carrier

A radio frequency modulated to carry information.

Channel

A frequency band in which a specific signal is transmitted. TV signals require a 6 MHz-wide band.

Coaxial Cable

A cable used to transmit high-frequency signals. RG-6 coaxial cable, for example, is used between the LNB and receiver.

DBS

Direct Broadcast Satellite, a new method of program delivery by which signals are sent directly to small (18-inch to 3-feet diameter) home dishes from high-powered (120- to 200-watts per transponder) Ku-band satellites, as opposed to "over the air" broadcast, cable delivery or lower-powered C-band or Ku-band transmissions. PrimeStar is a quasi-DBS service, while USSB and DirecTv, planned for 1994, and EchoStar, planned for 1995, will be true DBS services.

Decoder
A device that restores a signal to its original form after it has been encoded. Also called a descrambler or a decryption device.

Demodulator
A device that extracts signals from transmitted carrier waves.

Digital Audio
Method used to transmit audio on scrambled channels.

Dish
Nickname for a parabolic satellite antenna.

DNR
Dynamic Noise Reduction is a filter circuit that reduces high audio frequencies such as hiss.

Dolby Surround Sound
Four-channel audio format, which is encoded in the two audio channels of virtually every motion picture, music video and movie on television.

Downconverter
A circuit that lowers the high-frequency signal to a lower, intermediate range. The three types of downconversion are single, dual and block downconversion.

Downlink
Term used to describe the retransmitting of signals from a satellite back to Earth.

Dual Feedhorn
A feedhorn that can simultaneously receive both horizontally and vertically polarized signals.

Dual Band Feedhorn
A feedhorn that can receive both C-band and Ku-band signals.

Earth Station
Term used to describe a system for receiving signals.

EIRP
Abbreviation for effective isotropic radiated power. The combined result of transmitter or transponder RF power and transmitting antenna gain.

Elevation Angle
The vertical angle measured from the horizon to a targeted satellite.

Favorite Channel Memory
IRD feature that allows the storage of favorite satellite TV and ra-

dio stations in memory as "favorite channels" for easy recall.

Feedhorn

A device that gathers microwave signals reflected from the surface of the dish and feeds them to the LNB.

Footprint

The geographic area toward which a satellite directs its signal.

Frequency

The property of an alternating-current signal measured in cycles per second or hertz.

Geosynchronous

Term applied to satellites parked in orbit in the Clarke Belt (named for Arthur C. Clarke, father of satellite TV) 22,300 miles above the equator. Geostationary means same thing; often used interchangeably with geosynchronous.

Gigahertz (GHz)

One billion cycles per second. Signals above one gigahertz are known as microwaves.

Hertz (Hz)

Cycles per second.

Horizon-to-Horizon

Type of antenna mount that permits 180 degrees of antenna movement. Very strong and reliable, it is the most accurate mount to use for tracking Ku-band satellites.

IRD

Abbreviation for integrated receiver-decoder. New satellite receivers feature a built-in decoder (VideoCipher II Plus or VideoCipher RS) and dish motor drive controller. In older models, these were separate components.

Ku-band (also see C-band)

The microwave frequency band between approximately 11 and 13 GHz; an alternate satellite TV transmission system with up to 32 channels per satellite available to home viewers.

Low Noise Amplifier (LNA)

A device that receives and amplifies the weak signals reflected by the antenna via the feed.

Low Noise Block Downconverter (LNB)

A device that amplifies and downconverts the whole 500 MHz satellite bandwidth at once to an intermediate frequency range.

Megahertz (MHz)

Millions of cycles per second.

Microwave

See Gigahertz.

Modulation

The process in which a message is added to a carrier wave.

Modulator

A device that modulates a carrier. In a satellite receiver or VCR, the modulator places the audio and video signals on a carrier, usually channel 3 or 4, so they can be tuned by a standard TV set.

Mount

Structure that supports the antenna. It may be fixed, Az-El or polar.

MTS Stereo

Broadcast standard by which local stations transmit stereo programs to TVs and VCRs equipped with MTS tuners. Some new IRDs feature MTS output, permitting stereo programs to be delivered on a single cable to MTS TVs and VCRs throughout the house.

On-Screen Graphics

Displays on the TV screen that provide information such as satellite name and channel. Advanced IRD models are "menu-driven," permitting commands to be selected from easy-to-read lists of options.

Parental Control

Satellite receiver feature that allows restriction of viewing by locking out selected channels.

PIP

Picture-in-picture is a feature that allows display of a second signal source, such as that from a VCR or videodisc player, in a corner of the main screen.

Polar Mount

An antenna mount that permits all satellites to be scanned with movement of one axis.

Polarization

A technique used to increase the capacity of the satellite channels. In present C- and Ku-band systems, electromagnetic waves are polarized horizontally and vertically in order to reuse the same frequencies. Circular polarization schemes are used in many countries.

Private Terminal

A television receive-only earth terminal; a home dish system.

Programmer

A supplier of television programs transmitted via satellite such as HBO, TNT, WTBS, ESPN, etc.

Programmable Timer

IRD feature that allows the automatic selection of satellite and channel for programs to record on a VCR, even when user is asleep or not at home.

Satellite Receiver

A wideband FM (frequency modulation) receiver operating in the microwave range that detects baseband signals; part of an IRD.

Scrambling

Encryption or encoding process used to prevent viewing of subscription programs without authorization.

Skew

Term used to describe the fine-tuning adjustment of the feedhorn polarity probe.

Sparklies

Black and white dots that appear in a satellite-received picture as a result of a poor signal.

Teletext

An "electronic newspaper" transmitted via satellite that can be received with a teletext decoder such as the InfoCipher.

Transponder

A combination receiver, transmitter and antenna package on a satellite. The signal is received from the uplink station and downlinked back to Earth on a different frequency. Most C-band satellites have 24 transponders (12 vertical, 12 horizontal polarization); most Ku-band satellites have 16, although the Canadian Anik birds have 32. Some satellites have a mix of C-band and Ku-band transponders.

Tuner

The portion of a receiver that selects a desired signal from a group of signals in a frequency band.

TVRO

Trade name often used for satellite receiving system, shorthand for television receive-only. Home dish owners receive signals while programmers send or uplink signals.

Uplink

Antenna facility that transmits signals to a satellite.

Video Compression

A new, soon to be used technology that permits the squeezing of multiple channels into the bandwidth of one satellite transponder.

Video Monitor

A television set that accepts direct video signals. Not all TV sets are also monitors, which are best suited to deliver the most optimal video and audio performance.

Video Signal

That portion of the transmitted signal containing the picture information.

APPENDIX II
Troubleshooting

Preventative maintenance such as periodic lubrication of mechanical parts and weatherproofing outside connections will prolong the life expectancy of certain components and keep others from getting damaged by water and ice. But electronic devices do fail and mechanical parts wear out. Eventually, all satellite systems will require some sort of repair or replacement parts.

Your knowledge of the various system components and their functions, along with a methodical approach, can help isolate operational problems when they do occur. Oftentimes a thorough visual inspection will reveal something as simple as a broken wire, corroded connector or loose hardware on the mount. Problems such as these usually can be corrected without the need for a service call.

When it is necessary to call for help, your ability to explain the symptoms in an intelligent manner will assist the dealer in making an initial diagnosis and help him be prepared with the proper equipment to isolate and repair the problem quickly and efficiently.

Troubleshooting operational problems can be simplified by breaking the system down to three basic subsystems: mechanical, electromechanical and RF. The mechanical system consists of the mount, antenna, feed support and feedhorn. Actuators and servo motors have both electronic and mechanical parts. The LNB, cable, satellite receiver, channel 3/4 modulator and TV all play a part in processing the RF signals.

The first step in isolating a problem is to determine what is working and what is not working. That means doing some basic detective work. Here are some examples:

- Do the front panel lights come on? If not, maybe the receiver is unplugged or a fuse has blown.

• Do you get both even- and odd-numbered channels? If not, a wire may be loose or the servo motor might have failed.

• Can you move the dish from one satellite to another?

• Are you able to receive some satellites and not others? Did the problem coincide with a recent lightning storm or heavy winds? Are you still receiving signals from your local antenna?

The answers to questions such as these will help to eliminate certain components as the possible cause of the problem. If you're getting good pictures on all channels from one satellite but not others, you can eliminate the LNB, coaxial cable and receiver as possible suspects.

Let's suppose the receiver lights up and the dish moves, but there's no visible picture on any satellite. This symptom could be caused by a blown fuse, severed cable, bad LNB or inability of the dish to target a satellite. If the picture was lost during a lightning storm, a blown fuse or bad LNB would be likely causes. Pictures lost during a windstorm is an indication that the mount may have moved on the pole.

Start with a visual inspection. Make sure the LNB cable is properly connected at the receiver's input. Unplug the power before disconnecting or reconnecting the cable. Look to see that the center conductor is extended far enough out from the "F" connector to ensure a good electrical contact but not too far to cause a short inside the receiver. Also check the cable connection at the LNB for possible water intrusion or corrosion.

In the chapter on mounts we discussed how to use an inclinometer or Arc-Set tools to set the axis elevation and declination angles. With the dish actuated up to the zenith position (centered over the mount), verify these angles.

Sometimes a heavy wind can loosen hardware and cause the declination to slip. Wind also can cause the mount to rotate on the ground pole. In either case, the dish will not be able to accurately target satellites. If you scribed a line on the mount cap and pole as suggested, realigning the two marks should put the dish back on track.

The troubleshooting procedures outlined above often will result in finding the cause for a loss of picture on all satellites. However, further investigation will require the use of test equipment or substitution of components. Testing microwave components and repairing satellite receivers should be left to a competent technician. But simple voltage and continuity checks can be accomplished by someone with a basic knowledge of electronics and the use of a volt/ohm meter.

All block (950-1450 MHz input) receivers apply 18 volts to the LNB cable. This 18-volt supply usually is protected by either a circuit breaker or fuse. A volt/ohm meter (VOM) or digital volt meter (DVM) can be used to check the presence of voltage at the receiver input. With the negative probe of the meter connected to the chassis (ground) and the positive probe connected to the center pin on the female "F" fitting, you should measure about 18 volts on the DC scale. Zero voltage would indicate a blown fuse, tripped breaker or other power supply problem.

If replacing a blown fuse or resetting a breaker restores voltage at the input connector you will need to investigate the cable for possible shorts before reconnecting it.

With the LNB cable disconnected at the receiver and LNB, put the meter on the highest resistance scale and measure between the center conductor and the shield. Any resistance is an indication that either water has entered the cable or there is a direct short. Water sometimes can be evaporated with a hair dryer.

A broken LNB cable also will cause a loss of picture. Once you've checked the cable for shorts, test it for continuity by shorting the center conductor to the shield at one end and checking for resistance between the center conductor and shield at the other end. The meter should read zero or very low resistance. A high or infinite resistance would indicate a damaged or broken cable.

If the cable checks out okay and you have voltage at the receiver's input, connect the LNB cable to the receiver and measure the voltage between the shield and center conductor at the LNB end. You should measure 18 volts. If there is no picture when the cable is reconnected to the LNB or the fuse blows again, the LNB is bad.

LNBs can produce various symptoms, depending on the fault. A stage of amplification or the regulator can short causing the device to draw excessive current and blow fuses. Sometimes a stage of amplification will open and cause sparklies in the picture. Other than using a spectrum analyzer or measuring current draw, substitution is the best method of troubleshooting an LNB as a possible cause of picture loss or degradation.

Most feedhorns use a servo motor which moves a probe (the actual antenna) inside the throat to select either vertically or horizontally polarized signals. The latest LNBFs have no moving parts. Instead, polarity is switched electronically between two probes that are fixed. Servo-controlled feeds are another source of system problems, especially in areas subject to lightning storms.

Troubleshooting polarity problems is fairly simple. The servo motor is connected to the output of the receiver by three wires. One wire supplies five volts to power the small motor and another wire delivers a pulse to tell the servo motor when to change the probe from vertical to horizontal or vice versa. The third wire is ground. Whether the probe selects vertical signals on odd or even channels depends on the polarity format of the satellite being viewed. This information is usually stored into the receiver's memory on an on-satellite basis. Proper polarity or skew position then becomes automatic when changing channels and satellites.

If you can only receive odd-numbered channels on one satellite and even-numbered channels on another, it means the polarity is not changing. Something could be physically blocking the probe movement such as a wasp nest up inside the throat, or a wire has become disconnected, or the servo motor is frozen or burned out or the servo drive circuit has failed in the receiver.

Like any problem, the first step is visual inspection. Look for the obvious. Are the three servo wires connected at the receiver and at the feedhorn? Since the servo wires and LNB cables must run up the dish, there has to be enough slack behind the dish to permit dish movement without snagging the wires. All feedhorns should include a plastic cap over the throat to prevent wasps and small birds from building their homes inside.

You can see the probe by removing the plastic cap and looking up inside the feed. Have someone change channels while you observe the probe's movement. It should move 90 degrees back and forth. If you hear a buzz from the motor (not a wasp) but there's little or no probe movement, suspect a broken or worn servo motor.

If you hear no sound at all from the servo motor when channels are being changed inside, it's probably a bad connection or broken wire. You can use the same procedure to check the wires for shorts and continuity with a volt/ohm meter as you did for checking the LNB cable. Failure of the servo drive circuits in receivers is rare.

The actuator arm and motor which pushes and pulls the dish about its axis is the hardest working component in a satellite system and the cause for most service calls. Because it is a mechanical device and subjected to the elements, repair or replacement is common. In the chapter on actuators we explored the function of the sensor, problems related to water intrusion and preventative maintenance.

The sensor's job is to deliver pulses back to the controller which keeps track of the dish position. When pulses are not received the controller will shut down and usually display some sort of error. This symptom can be caused by a bad sensor or connection, broken wire, frozen jack or bad motor.

Troubleshooting a bad sensor is simple with a volt/ohm meter. Reed sensors are switches which respond to a magnetic field. Every time one of the magnets passes in front of the switch it closes and then opens as the magnet is moved away. All you have to do is disconnect the switch and connect the two wires to the meter. While observing the meter movement on a resistance setting, pass a magnet within close proximity to the switch. You should see the meter indicate a short when the magnet is near the switch and an open circuit when the magnet is moved away.

Other problems related to sensors include loose or cracked magnets, or a loose magnet wheel which can cause intermittent pulses. Similar problems can result if the sensor is not mounted at the proper distance (about the thickness of a matchbook cover) from the magnets.

Some motor drives are protected with a clutch that is designed to slip if the jack freezes or fails to move for any reason. When this happens the motor continues to turn and a clicking sound can be heard. It's a feature built in to prevent the motor from stripping gears. Since the magnet wheel turns with the motor, pulses will continue to the controller even though the dish may not be moving. This often happens with this type of actuator when someone overrides the programmable limits in the controller and runs the arm up against its mechanical limits of travel. The result is that the actual dish position gets out of synchronization with the programmed positions in the controller. Most IRDs have a feature to resync the controller memory with the position of the arm without the need to reprogram every satellite location.

The sensor is not always the culprit when a dish fails to move. When the original lubricants in an actuator arm dry up or get displaced by water the arm can bind. Prior to this condition you usually can hear a grinding sound when the arm is extended or retracted. Motors also can fill with water. Water intrusion always results in corrosion, which has a damaging effect on armatures, brushes and bushings. Protective covers and drain holes will help extend the life of an actuator and motor. But service and/or replacement is inevitable.

In cases where the sensor checks out, but the motor fails to extend or retract the arm, you can determine which is the culprit by separating the motor from the arm. If the motor runs normally when not coupled to the arm, the telescoping arm is probably frozen. In either case, it's best to replace the entire unit.

A surprising amount of system difficulties are traced to pilot error. This is especially true with complex systems that include VCRs, audio/video amplifiers, RF switches and other signal distribution devices. Often, the inability to tune satellite channels results because someone left a switch in a position to monitor the VCR or outside antenna. In other cases the culprit may be something as simple as a two-way splitter.

When troubleshooting problems in more complex entertainment systems, use a systematic approach, eliminating one component, switch or splitter at a time.

The key to successful troubleshooting of any satellite system, however, is understanding the basic principles of how each subsystem works and how they interact with each other. That has been the goal of *All About Satellite TV*.

If you have carefully read each chapter in this book, you should be able to troubleshoot your own system. Ideally, you should have learned enough to tell a professional repair technician what the trouble is and where to look for its cause. Who knows? You might even have learned enough to fix some problems yourself. If that's the case, then *All About Satellite TV* has more than accomplished its goal of being a basic introduction and guide.

APPENDIX III

Factory Bulletins

Today's IRDs represent the lastest technologies which are the result of extensive research and development. Prototypes are exposed to extreme temperatures, simulated lightning strikes, power surges and numerous other stress tests prior to production. In addition, new products are continually tested to ensure reliability before they are shipped to distributors.

Although rigorous testing and quality control help to reduce failure rates, IRDs occasionally fail and require service. When this happens, you'll need to call a dealer or send the unit to an authorized service center for repair. There are no user-servicable parts inside an IRD. Fuses and circuit breakers are usually accessible on the rear panel.

IRD manufacturers are always looking for ways to improve their products and make them more reliable. This applies to older models as well as current ones. If a company finds a high failure rate with a certain component or circuit design, it often will generate a service bulletin to its authorized service centers that outlines the procedures necessary to correct the problem. One advantage of sending a unit to an authorized service center is that the upgrades and modifications generally will be incorporated along with the repair.

Upgrades to older models may include the replacement of a component because it had a high failure rate or changing the value of a component to improve performance. In some cases, it may mean changing an entire circuit board or power supply. Software upgrades also are available for some older IRDs and receivers. Preprogrammed information such as satellite formats, favorite channel lists and audio frequencies are prestored in many IRDs. When programmers move from one channel or satellite to another these parameters must be "written over" to reflect those changes. Some manufactures offer plug-in EPROMs which will bring the software up to date with the latest changes, including new satellite formats and favorite program lists.

Factory service and technical bulletins are considered proprietary information and usually are not available to consumers. You'll need to contact your dealer to find out if any upgrades in hardware or software are available for your particular equipment.

—— Factory Service Centers ——

The following is a list of factory service centers. Be sure to call ahead before shipping products for repair. Some service centers require a return authorization number.

Chaparral Communications	2450 North First St. San Jose, Calif. 95131	408-435-3088
Channel Master	Industrial Park Dr. Smithfield, N.C. 27577	919-989-2211
Drake Service (R.L. Drake)	230 Industrial Dr. Franklin, Ohio 45005	513-746-6990
DX Communications	10 Skyline Dr. Hawthorne, N.Y. 10532	914-347-4040
Echosphere Corporation	90 Inverness Circle E. Englewood, Colo. 80015	303-790-7878
Fujitsu General	353 Rt. 46 W Fairfield, N.J. 07004	201-575-0380
General Instrument	Check back of warranty card	704-327-2026
HTS (Houston Tracker Systems)	90 Inverness Circle East Englewood, Colo. 80015	303-790-7878
Panasonic Industrial Service	4245 International Blvd., Suite B Norcross, Ga. 30093	404-717-6855 1-800-524-1448 (Tech Service)
Satellite Service Company (STS)	1409 Washington Ave. St. Louis, Mo. 63103	314-421-0102
Tee-Comm (Startrak)	7016 N.W. 50th Miami, Fl. 33166	305-477-3298
Toshiba America	1010 Johnson Dr. Buffalo Grove, Il. 60089	708-541-9400 (Ext. 370)
Uniden Service Center	8707 North by Northeast Blvd. Indianapolis, Ind. 46250	317-842-2483 (Ext. 501)
Zenith Electronics Corp.	230 Industrial Dr. Franklin, Ohio 45005	513-746-6990

Independent Satellite Service Centers

The following is a list of independent service companies that repair most makes and models of IRDs, receivers and positioners.

Birdview Service Center	1407 E. Spruce Olathe, Kan. 66061	913-829-6240
DigiComm Electronics, Inc.	22882 Pontiac Trail South Lyon, Mich. 48178	313-486-4343
Pacific Satellite and Electronics	1401 Lincoln Ave. Napa, Calif. 94558	707-226-7488
Protronics	8 Raven's Pointe Dr. Lake St. Louis, Mo. 63367	314-625-1022
PTS	5233 Hwy. 37 South Bloomington, Ind. 47401	812-824-9331
Satellite TV Service Center	736-B Tunnel Rd. Asheville, N.C. 28805	704-298-1448
Quality Controlled Electronics	38424 Webb Dr. Westland, Mich. 48185	313-721-5666

Index